Holt Mathematics

Chapter 10 Resource Book

HOLT, RINEHART AND WINSTON
A Harcourt Education Company
Orlando • Austin • New York • San Diego • London

Printed in the United States of America

ISBN 0-03-078401-8

6 7 8 170 09 08

CONTENTS

Blackline Masters

Date ___________

Dear Family,

In this chapter, your child will learn about probability: how to work with experimental probability, theoretical probability, and the Fundamental Counting Principle. Probability will be related to insurance rates, genetics, sports, games, and the chances of disasters such as earthquakes occurring.

You can find the probability of an event by using the definition of probability.

- A probability of 0 means the event is **impossible,** or can never happen.
- A probability of 1 means the event is **certain** or has to happen.
- The probabilities of all the outcomes add up to 1.

Never happens		Happens about half the time		Always happens
0	$\frac{1}{4}$	$\frac{1}{2}$	$\frac{3}{4}$	0
0	0.25	0.50	0.75	1
0%	25%	50%	75%	100%

You can find the probabilities of outcomes in a sample space by giving the probability of each outcome. **The weather forecast shows a 30% chance of snow.** The probability of rain is ***P*(snow) = 30% = 0.3.** The probabilities must add to 1, so the probability of no snow is ***P*(no snow) = 1 – 0.3 = 0.7, or 70%.**

Outcome	Snow	No snow
Probability	0.3	0.7

You can estimate the probability of an event by using experimental methods. After 1000 spins of the spinner, the following information was recorded. Estimate the probability of the spinner landing on red.

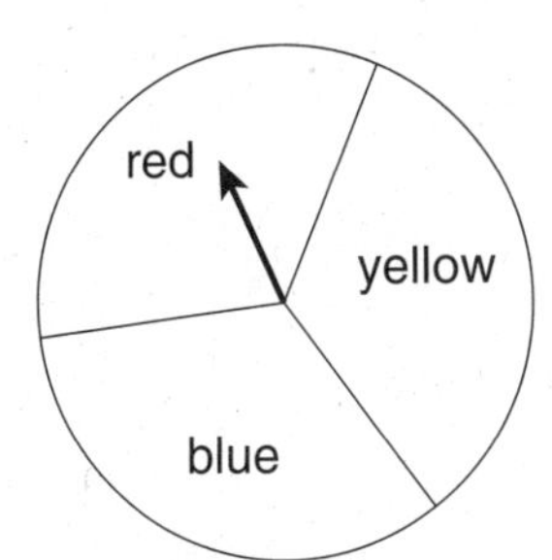

Outcome	Blue	Red	Yellow
Spins	448	267	285

$$\text{probability} \approx \frac{\text{number of spins that land on red}}{\text{total number of spins}} = \frac{267}{1000} = 0.267$$

The probability of spinning red is about 0.267, or 26.7%.

Theoretical probability is used to estimate probabilities by making certain assumptions about an experiment. The probabilities must add to 1.

A coin, number cube, or other object is called **fair** if all outcomes are equally likely. To calculate a **theoretical probability,** an experiment consists of rolling a fair number cube. There are 6 possible outcomes: 1, 2, 3, 4, 5, and 6.

What is the probability of rolling a 3?
The number cube is fair; so the 6 outcomes are equally likely.
The probability of rolling a 3 is $P(3) = \frac{1}{6}$.

The Fundamental Counting Principle
If there are *m* ways to choose a first item, and *n* ways to choose a second item after the first item has been chosen, then there are $m \cdot n$ ways to choose all items.

Your child will learn to find the number of possible outcomes in an experiment and how to use the **Fundamental Counting Principle.**

Find the number of possible new 7-digit phone numbers for a new area code. All phone numbers are equally likely.

First digit	Second digit	Third digit	Fourth digit	Fifth digit	Sixth digit	Seventh digit
?	?	?	?	?	?	?
10 choices	10 choices	10 choices	10 choices	10 choices	10 choices	10 choices

$10 \cdot 10 \cdot 10 \cdot 10 \cdot 10 \cdot 10 \cdot 10 \cdot = 10{,}000{,}000$.

The number of possible 7-digit phone numbers is 10,000,000.

For additional resources, visit go.hrw.com and enter the keyword MT7 Parent.

Name ______________________ Date ____________ Class ____________

Practice A
Probability

1. Meteorology is the study of the atmosphere, natural phenomenon, atmospheric conditions, weather and climate. A meteorologist forecasts and reports the weather. A meteorologist forecasts a 70% chance of rain. What is the probability of each outcome?

Outcome	Rain	No rain
Probability		

Use the spinner to determine the probability of each outcome.

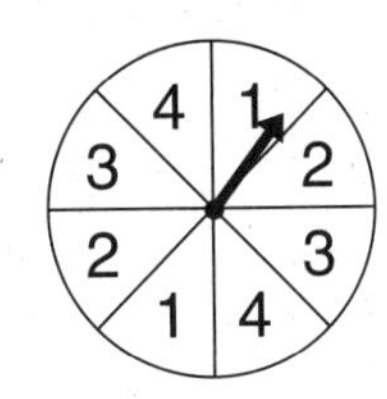

2. $P(1)$ ____________

3. $P(4)$ ____________

4. $P(\text{even number})$ ____________

5. $P(5)$ ____________

6. $P(\text{odd number})$ ____________

7. $P(\text{an integer})$ ____________

8. $P(1 \text{ or } 2)$ ____________

9. $P(\text{number} > 1)$ ____________

10. $P(\text{a whole number})$ ____________

11. Mrs. Silverstein has 14 boys in her class of 25 students. She must select one student at random to serve as the class moderator. What is the probability that she will choose a boy? What is the probability that she will choose a girl?

____

12. When tossing a regular coin, what is the probability of it landing on heads?

____

Name ______________________ Date ____________ Class ____________

LESSON 10-1

Practice B

Probability

These are the results of the last math test. The teacher determines that anyone with a grade of more than 70 passed the test. Give the probability for the indicated grade.

Grade	65	70	80	90	100
# of Students	5	3	12	10	2

1. $P(70)$ ____________
2. $P(100)$ ____________
3. $P(80)$ ____________
4. $P(\text{passing})$ ____________
5. $P(\text{grade} > 80)$ ____________
6. $P(60)$ ____________
7. $P(\text{failing})$ ____________
8. $P(\text{grade} \leq 80)$ ____________

A bowling game consists of rolling a ball and knocking up to 5 pins down. The number of pins knocked down are then counted. The table gives the probability of each outcome.

Number of Pins Down	0	1	2	3	4	5
Probability	0.175	0.189	0.264	0.205	0.132	0.035

9. What is the probability of knocking down all 5 pins?

____

10. What is the probability of knocking down no pins?

____

11. What is the probability of knocking down at most 2 pins?

____

12. What is the probability of knocking down at least 2 pins?

____

13. What is the probability of knocking down more than 3 pins?

____

Name ______________________ Date ___________ Class ___________

LESSON 10-1

Practice C

Probability

Demographers often use statistics to predict and explain future changes in populations in many areas including housing, education, life events, and unemployment. A demographer developed this chart to illustrate the cause of death in his community of 50,000 people.

Give the probability for each outcome.

Outcome	Heart	Cancer	Accident	Respiratory	Other
Probablility	0.35	0.28	0.16	0.13	0.08

1. *P*(death from cancer) ___________

2. *P*(death from accident) ___________

3. *P*(death from heart) ___________

4. *P*(non-accidental death) ___________

5. *P*(death from other) ___________

6. *P*(death from heart or cancer) ___________

Use the spinner to determine the probability of each outcome.

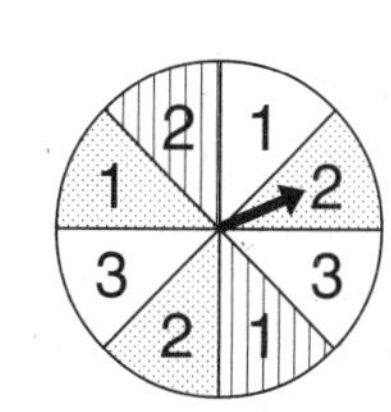

7. *P*(white 1) ___________

8. *P*(dots 2) ___________

9. *P*(lines even) ___________

10. *P*(dots 1) ___________

11. *P*(white odd) ___________

12. *P*(dots integer) ___________

13. *P*(odd) ___________

14. *P*(white or 2) ___________

15. There are six teams competing to collect the most food for the food bank. Team B has a 30% chance of winning. Teams A, C, D, and E all have the same chance of winning. Team F is one third as likely to win as Team B. Create a table of probabilities for the sample space.

Outcome						
Probablility						

Name ______________________ Date __________ Class __________

LESSON
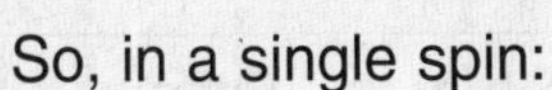

Reteach

10-1 *Probability*

The **probability** that something will happen is how often you can expect that **event** to occur. This depends upon how many outcomes are possible, the **sample space.**

In the spinner shown, the circle is divided into four equal parts. There are 4 possible outcomes.

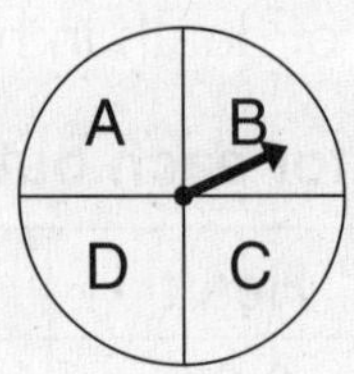

So, in a single spin:

$P(A) = P(B) = P(C) = P(D) = 25\% = \frac{1}{4}$

Complete to give the probability for each event.

	1. A fair coin is tossed.	**2.** A number cube is rolled.
List all the possible outcomes.	________ ________	________ ________
How many outcomes in sample space?	________	________
Find the probability of the event shown.	*P*(heads) = _____	*P*(5) = _____

- A probability of 0 means the event is **impossible,** or can never happen.
  On the spinner above, *P(F)* = 0.
- A probability of 1 means the event is **certain,** or has to happen.
  In one roll of a number cube, *P*(a whole number from 1 through 6) = 1.

Give the probability for each event.

3. selecting a rectangle from the set of squares

P(rectangle) = _____

4. selecting a negative number from the set of whole numbers

P(negative number) = _____

- The sum of the probabilities of all the possible outcomes in a sample space is 1.
  If the probability of *snow* is 30%, then the probability of *no snow* is 70%.
  P(snow) + *P*(no snow) = 1

5. If the probability of selecting a senior for a committee is 60%, then the probability of not selecting a senior is:

____

6. If the probability of choosing a red ball from a certain box is 0.35, then the probability of not choosing a red ball is:

____

Name ______________________ Date __________ Class __________

LESSON **10-1**

Reteach

Probability (continued)

To find the probability that an event will occur, add the probabilities of all the outcomes included in the event.

This bar graph shows the midterm grades of the 30 students in Ms. Lin's class.

What is the probability that Susan has a grade of C or higher?

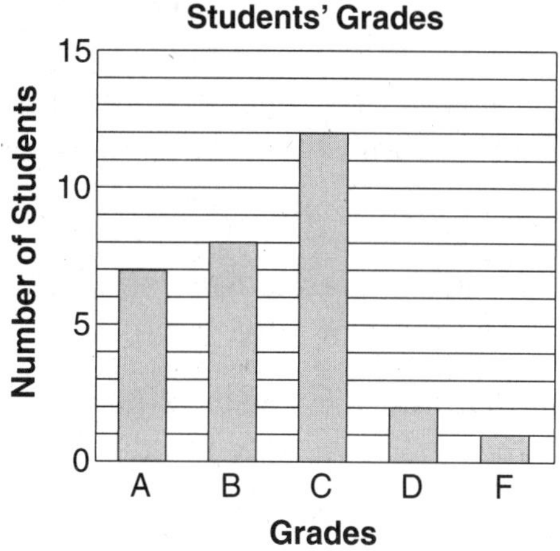

The event "a grade of C or higher" consists of the outcomes C, B, A.

$P(\text{C or higher}) = P(\text{C}) + P(\text{B}) + P(\text{A})$

$$= \frac{12}{30} + \frac{8}{30} + \frac{7}{30} = \frac{12 + 8 + 7}{30} = \frac{27}{30} = \frac{9}{10} = 90\%$$

So, the probability that Susan's midterm grade is C or higher is $\frac{9}{10}$ or 90%.

Use the bar graph above to find each probability.

7. B or higher is honor roll. What is the probability that Ken made the honor roll?

$P(\text{B or higher}) = P(\text{B}) +$ ____

$= \frac{\_\_}{30} + \frac{\_\_}{30}$

$= \frac{\_\_}{30} =$ ____

So, the probability that Ken made the honor roll is: ____ or ____%.

8. In this class, D is a failing grade. What is the probability that Tom failed?

$P(\text{D or lower}) = P(\text{D}) +$ ____

$= \frac{\_\_}{30} + \frac{\_\_}{30}$

$= \frac{\_\_}{30} =$ ____

So, the probability that Tom failed is: ____ or ____%.

Name ______________________ Date ____________ Class ____________

LESSON 10-1

Challenge
Why We Look Like Our Parents

Each parent carries two genes with respect to a specific trait and each passes one of these genes on to an offspring who then also has two genes for that trait.

In pea plants, a tall gene is dominant over a short gene. So, if a pea plant has at least one tall gene, the plant is tall. If T represents *tall* and t represents *short*, one way to represent the gene makeup with respect to height of a tall pea plant would be Tt.

1. What is another way to represent the gene makeup with respect to height of a tall pea plant? ____________

An early 20th-century English geneticist, Reginald Punnett, invented a method to display the gene makeup of parents and their offspring.

2. The **Punnett square** at the right shows the gene makeup of one tall parent plant as the labels for the columns. Insert your result from Question 1 for the other tall parent plant as the row labels.

	T	t
___		
___		

3. a. The label for each column has been inserted in each box of its column, as the first gene of the offspring plant. Insert your labels for each row as the second gene of each offspring plant.

b. According to the two genes now in each of the boxes for the new offspring plants, tell if the new plant will be tall or short.

c. What is the probability that an offspring of these tall parent plants will be tall?

____

	T	t
T	T	t
T	T	t

4. Suppose the gene makeup for both tall parent pea plants is Tt.

a. Complete a Punnett square to display the gene makeup of the offspring.

b. What is the probability that an offspring of these tall parent plants will be tall?

____

Name ______________________ Date __________ Class __________

LESSON 10-1

Problem Solving

Probability

Write the correct answer.

1. To get people to buy more of their product, a company advertises that in selected boxes of their popsicles is a super hero trading card. There is a $\frac{1}{4}$ chance of getting a trading card in a box. What is the probability that there will not be a trading card in the box of popsicles that you buy?

   ____

2. The probability of winning a lucky wheel television game show in which 6 preselected numbers are spun on a wheel numbered 1–49 is $\frac{1}{13{,}983{,}816}$ or 0.000007151%. What is the probability that you will not win the game show?

   ____

Based on world statistics, the probability of identical twins is 0.004, while the probability of fraternal twins is 0.023.

3. What is the probability that a person chosen at random from the world will be a twin?

   ____

4. What is the probability that a person chosen at random from the world will not be a twin?

   ____

Use the table below that shows the probability of multiple births by country. Choose the letter for the best answer.

5. In which country is it most likely to have multiple births?

   A Japan
   B United States
   C Sweden
   D Switzerland

6. In which country is it least likely to have multiple births?

   F Japan
   G United States
   H Sweden
   J Switzerland

7. In which two countries are multiple births equally likely?

   A United Kingdom, Canada
   B Canada, Switzerland
   C Sweden, United Kingdom
   D Japan, United States

Probability of Multiple Births

Country	Probability
Canada	0.012
Japan	0.008
United Kingdom	0.014
United States	0.029
Sweden	0.014
Switzerland	0.013

Name ______________________ Date ____________ Class ____________

LESSON 10-1

Reading Strategies

Focus On Vocabulary

Probability is the chance that something will happen.

An **event** has an outcome that can be stated using probability. The probability of something happening is written ***P*(event).**

The probability of an event happening is described by a number between 0 and 1, or as a percent between 0% and 100%.

A probability of 0 or 0% means it is **impossible** for the event to happen.

A probability of 1 or 100% means it is **certain** the event will happen.

Answer each question.

1. What name is given to the chance of something happening?

____

2. How is the probability of an event written?

____

3. What number or percent is used to mean that an event is certain?

____

An **experiment** is an activity where results are observed. Flipping a coin and tossing a number cube are both experiments.

In an experiment, each observation (such as one coin toss) is called a **trial.** The result of the trial is called an **outcome** (such as a coin landing on heads).

Use the terms above to answer the following questions.

4. What do you call an activity in which results are observed?

____

5. What is one observation in an experiment called?

____

6. What is an outcome?

____

Name ________________ Date ________ Class ________

LESSON 10-1

Puzzles, Twisters & Teasers

What Are Your Chances?

Solve the crossword puzzle.

Across

3. A probability of 0 means the event is ___.

5. The probabilities of all the outcomes in the sample space add up to ___.

7. An ___ is any set of one or more outcomes.

9. Each observation in an experiment is called a ___.

Down

1. Each result of an experiment is called an ___.

2. A probability of 1 means an event is ___.

4. The ___ of an event is a number that tells how likely the event is to happen.

6. An ___ is an activity in which results are observed.

8. The set of all possible outcomes of an experiment is the ___ space.

Name ________________________ Date ____________ Class ____________

LESSON 10-2

Practice A

Experimental Probability

The results of an unbiased survey show the favorite instruments of 8th graders. Estimate the probability of each.

Result	Piano	Drums	Trombone	Flute	Violin	Clarinet
Number	1	4	42	38	12	3

1. a student chooses clarinet

____

2. a student chooses drums

____

3. a student chooses flute

____

4. a student chooses piano

____

5. a student chooses trombone

____

6. a student chooses violin

____

A can contains color chips in 5 different colors. Thomas took a sample from the can and counted the colors. His results are in the table below.

Color	Blue	Pink	Black	White	Green
Number	10	5	20	30	15

7. Use the table to compare the probability that Thomas chooses a pink color chip to the probability that he chooses a white color chip.

____

8. Use the table to compare the probability that Thomas chooses a green color chip to the probability that he chooses a blue color chip.

____

9. Cheryl surveyed 30 students who ride the bus to school, 8 who walk, 9 who ride bicycles, and 3 who ride in cars. Estimate the probability that the next student Cheryl surveys will walk to school.

____

Name ______________________ Date __________ Class __________

LESSON 10-2

Practice B

Experimental Probability

1. A number cube was thrown 150 times. The results are shown in the table below. Estimate the probability for each outcome.

Outcome	1	2	3	4	5	6
Frequency	33	21	15	36	27	18
Probability						

A movie theater sells popcorn in small, medium, large and jumbo sizes. The customers of the first show purchase 4 small, 20 medium, 40 large, and 16 jumbo containers of popcorn. Estimate the probability of the purchase of each of the different size containers of popcorn.

2. *P*(small container) ____________

3. *P*(medium container) ____________

4. *P*(large container) ____________

5. *P*(jumbo container) ____________

Janessa polled 154 students about their favorite winter sport.

Outcome	Frequency
Skiing	46
Sledding	21
Snowboarding	64
Ice Skating	14
Other	9

6. Use the table to compare the probability that a student chose snowboarding to the probability that a student chose skiing.

____

7. Use the table to compare the probability that a student chose ice skating to the probability that a student chose sledding.

____

8. The class president made 75 copies of the flyer advertising the school play. It was found that 8 of the copies were defective. Estimate the probability that a flyer will be printed properly. ____________

Name ____________________ Date __________ Class __________

LESSON 10-2

Practice C

Experimental Probability

The developer of a Web page wants to track the number of hits to each link of the Web page. An automatic counter records the following hits in one week: home, 60 hits; FAQ, 20 hits; employment opportunities, 15 hits; products, 50 hits; order status, 30 hits; and contact information, 25 hits. Estimate the probability of each.

1. *P*(home) ______________

2. *P*(FAQ) ______________

3. *P*(products) ______________

4. *P*(order status) ______________

5. *P*(employment opportunities) ______________%

6. *P*(contact information) ______________

Hayley bought a CD with 12 songs on it. She placed it in her CD changer and selected random play mode. Hayley kept a record of how the tracks played. The following table illustrates the results.

Track	1	2	3	4	5	6	7	8	9	10	11	12
Frequency	2	4	3	1	2	4	2	3	5	2	1	4

Estimate the probability for each of the following.

7. *P*(track 2) ______________

8. *P*(track 4) ______________

9. *P*(track 5) ______________

10. *P*(track 8) ______________

11. *P*(track 9) ______________

12. *P*(track 13) ______________

13. Use the table to compare the probability that Track 10 was played to the probability that Track 6 was played.

____

14. A coin is tossed 70 times, and it lands on heads 36 times. Estimate the probability of it landing on tails. ______________

Name ______________________ Date __________ Class __________

LESSON 10-2

Reteach

Experimental Probability

A machine is filling boxes of apples by choosing 50 apples at random from a selection of six types of apples. An inspector records the results for one filled box in the table below.

Type	Pink Lady	Red Delicious	Granny Smith	Golden Delicious	Fuji	MacIntosh
Number	8	12	6	4	15	5

The inspector then expands the table to find the experimental probability.

$$\text{probability} = \frac{\text{number of type of apple}}{\text{total number of apples}}$$

Type	Pink Lady	Red Delicious	Granny Smith	Golden Delicious	Fuji	MacIntosh
Experimental Probability (ratio)	$\frac{8}{50}$, or $\frac{4}{25}$	$\frac{12}{50}$, or $\frac{6}{25}$	$\frac{6}{50}$, or $\frac{3}{25}$	$\frac{4}{50}$, or $\frac{2}{25}$	$\frac{15}{50}$, or $\frac{3}{10}$	$\frac{5}{50}$, or $\frac{1}{10}$
Experimental Probability (percent)	16%	24%	12%	8%	30%	10%

Find each sum for the apple experiment.

1. The sum of the experimental probability ratios.

   $\text{probability} = \frac{8}{50} + \frac{12}{50} + \frac{6}{50} + \frac{4}{50} + \frac{15}{50} + \frac{5}{50} = \frac{\_\_}{50}$ or ____

2. The sum of the experimental probability percents.

   probability = 16% + 24% + 12% + 8% + 30% + 10% = ____% or ____

Complete the table to find the experimental probability.

3. Five types of seed are inserted at random in a pre-seeded strip ready for planting.

Type	Marigold	Impatiens	Snapdragon	Daisy	Petunia
Number	40	100	80	60	120
Experimental Probability (ratio)	$\frac{\_\_}{400}$, or ____	$\frac{\_\_}{400}$, or ____	$\frac{\_\_}{400}$, or ____	$\frac{\_\_}{400}$, or ____	$\frac{\_\_}{400}$, or ____
Experimental Probability (percent)					

Name ______________________ Date __________ Class __________

LESSON 10-2

Challenge

Tossing and Spinning

The more times you repeat an experiment, the closer the experimental probabiltiy and the theoretical probability become.

Toss a penny 200 times.

Heads	Tails

1. Record your results in the table.

2. What is the theoretical probability of:

getting heads? __________ getting tails? __________

3. What is your experimental probability of:

getting heads? __________ getting tails? __________

4. How close are your experimental probabilities to the theoretical probabilities?

____

Spin a penny 200 times.

Heads	Tails

5. Record your results in the table.

6. What is your experimental probability of:

getting heads? __________ getting tails? __________

7. Compare your experimental probabilities for tossing the penny and spinning the penny. Are they close? Explain.

____

Roll a number cube 200 times.

1	2	3	4	5	6

8. Record your results in the table.

9. What is the theoretical probability of:

getting a 1? _____ a 2? _____ a 3? _____ a 4? _____ a 5? _____ a 6? _____

10. What is your experimental probability of getting:

a 1? _____ a 2? _____ a 3? _____ a 4? _____ a 5? _____ a 6? _____

11. How close are your experimental probabilities to the theoretical probabilities?

____

Name ________________________ Date ____________ Class ____________

LESSON **10-2**

Problem Solving

Experimental Probability

Use the table below. Round to the nearest percent. Write the correct answer.

Average Number of Days of Sunshine Per Year for Selected Cities

City	Number of Days
Buffalo, NY	175
Fort Wayne, IN	215
Miami, FL	256
Raleigh, NC	212
Richmond, VA	230

1. Estimate the probability of sunshine in Buffalo, NY.

____

2. Estimate the probability of sunshine in Fort Wayne, IN.

____

3. Estimate the probability of sunshine in Miami, FL.

____

4. Estimate the probability that it will not be sunny in Raleigh, NC.

____

5. Estimate the probability that it will not be sunny in Miami, FL.

____

6. Estimate the probability of sunshine in Richmond, VA.

____

Use the table below that shows the number of deaths and injuries caused by lightning strikes. Choose the letter for the best answer.

States with Most Lightning Deaths

State	Average deaths per year	Average injuries per year	Population
Florida	9.6	32.7	15,982,378
North Carolina	4.6	12.9	8,049,313
Texas	4.6	9.3	20,851,820
New York	3.6	12.5	18,976,457
Tennessee	3.4	9.7	5,689,283

7. Estimate the probability of being injured by a lightning strike in New York.

   A 0.0000007%   **C** 0.00007%

   B 0.0000002%   **D** 0.000002%

8. Estimate the probability of being killed by lightning in North Carolina.

   F 0.0000006%   **H** 0.00002%

   G 0.00006%   **J** 0.000002%

9. Estimate the probability of being struck by lightning in Florida.

   A 0.00006%

   B 0.00026%

   C 0.0000026%

   D 0.0006%

10. In which two states do you have the highest probability of being struck by lightning?

    F Florida, North Carolina

    G Florida, Tennessee

    H Texas, New York

    J North Carolina, Tennessee

Name ________________________ Date ____________ Class ____________

LESSON 10-2

Reading Strategies

Make Predictions

Experimental probability is a statement of the results of a number of trials.

$$\text{Probability} = \frac{\text{number of times an event happens}}{\text{total number of trials}}$$

When you spin this spinner, it could land on the section with dots, the striped section, or the white section.

Use the spinner to answer the following questions.

1. Predict which section the spinner will land on most often. Why?

____

2. Predict which section the spinner will land on least often. Why?

____

The actual outcome of an experiment may or may not match your predictions. This chart shows the outcomes of 500 trials.

Outcome	Striped	Spotted	White
Spins	163	152	185

Answer the following questions.

3. How many times did the spinner land on the striped section? ________

4. Did *P*(striped) match your prediction for the outcome of the experiment? Explain.

____

____

5. Did *P*(spotted) match your prediction for the outcome of the experiment? Explain.

____

____

Name ______________________ Date __________ Class __________

LESSON 10-2

Puzzles, Twisters & Teasers

Probable Problems

Estimate the probability of drawing a yellow marble in each situation below. Use the letters to answer the riddle.

1. probability: ______% **E**

Outcome	Blue	Yellow	Green	Black
Draws	217	201	295	287

2. probability: ______% **V**

Outcome	Blue	Yellow	Green	White
Draws	33	13	46	8

3. probability: ______% **C**

Outcome	Blue	Yellow	Green	White	Red
Draws	20	17	32	18	13

4. probability: ______% **H**

Outcome	Blue	Red	Yellow	White	Green
Draws	30	18	18	21	13

5. probability: ______% **D**

Outcome	Blue	Yellow	Green	Black	White
Draws	17	11	25	28	19

6. probability: ______% **A**

Outcome	Blue	Yellow	Green	White	Red	Black
Draws	65	35	235	180	369	116

7. probability: ______% **N**

Outcome	Blue	Yellow	Green	Black	White	Brown
Draws	7	21	15	18	19	20

8. probability: ______% **W**

Outcome	Blue	Red	Yellow
Draws	448	267	285

9. probability: ______% **Y**

Outcome	Blue	Yellow	Green
Draws	330	233	437

10. probability: ______% **L**

Outcome	Blue	Yellow
Draws	65	35

What do Christmas and a cat on the beach have in common?

They both ___ ___ ___ ___ **SA** ___ ___ ___
18 3.5 13 20.1 21 11 23.3

___ ___ **A** ___ **S**
17 35 28.5

Name ______________________ Date __________ Class __________

LESSON 10-3

Practice A
Use a Simulation

1. At a local salad bar, 5 out of every 7 customers order Cobb salad. What is the probability to the nearest percent that a customer will order a Cobb salad? __________

Use the table of random numbers to simulate the situation for Exercises 2–4.

7	6	3	6	4	5	2	2	4	3
4	5	4	7	5	3	7	4	4	5
6	4	7	4	4	6	4	7	5	2
3	7	1	3	3	3	3	5	7	1
1	2	6	2	5	5	4	4	2	5
4	3	5	2	3	3	7	4	4	3
4	4	5	1	6	2	4	6	7	5
5	4	5	6	6	3	6	3	2	3
3	6	3	2	7	2	2	4	4	4
5	7	5	2	4	5	2	4	7	6

2. Let the numbers 1–5 represent people who order Cobb salad. Checking in rows, how many people would you have to survey before you find 5 people who ordered Cobb salad? __________

3. What is the probability represented by Exercise 2? __________

4. Of the 100 customers represented in the random number table, what is the probability to the nearest percent that the customer will order a Cobb salad? __________

5. The owner of the salad bar makes the Cobb salad the special of the week. This increases sales to 6 out of 7 customers ordering a Cobb salad. What is the probability to the nearest percent that a customer will order the special? __________

6. Use the numbers 1–6 in the random number table to represent the customers who purchased a Cobb salad. Of the 100 customers represented in the random number, what is the probability to the nearest percent that the customer will order the special Cobb salad? __________

Name ______________________ Date __________ Class __________

Practice B
LESSON 10-3 *Use a Simulation*

Use the table of random numbers for the problems below.

8125	4764	7693	3675	1642	7988	7048	9135	3138	3256
9566	4413	7215	7992	4320	7438	3805	5473	8847	2397
7336	5393	8623	8570	5095	5685	6695	3570	3605	4656
6470	6065	8239	2953	5942	6496	8899	0701	5368	2106
5210	2570	8137	3587	3578	6657	6636	7188	5717	1770
4329	4110	2655	8258	9928	3873	5609	3695	7091	0368
5315	2654	0484	4601	4336	6624	5403	5870	8545	3905
2361	9097	3753	2498	0544	0923	6099	1737	4025	1221
2677	7741	5342	9844	3722	5120	8742	1382	2842	7386
3292	5084	1130	2747	0664	9718	6072	9432	7008	2024

Mr. Domino gave the same math test to all three of his math classes. In the first two classes, 80% of the students passed the test. If the third class has 20 students, estimate the number of students who will pass the test.

1. Using the first row as the first trial, count the successful outcomes and name the unsuccessful outcomes.

____

2. Count and name the successful outcomes in the second row as the second trial.

____

Determine the successful outcomes in the remaining rows of the random number table.

3. third row ____________

4. fourth row ____________

5. fifth row ____________

6. sixth row ____________

7. seventh row ____________

8. eighth row ____________

9. ninth row ____________

10. tenth row ____________

11. Based on the simulation, estimate the probability that 80% of the class will pass the math test. ____________

Name ______________________ Date ____________ Class ____________

LESSON 10-3

Practice C

Use a Simulation

Use the table of random numbers to simulate each situation. Use at least 10 trials for each simulation.

27768	56420	77775	39422	60423	71178	54012	21367
20182	54386	85157	89029	26369	14161	82065	86070
36558	13616	68098	21724	29916	78974	29433	52156
79405	19383	84186	06775	48080	31018	91551	25107
29426	00966	47941	68043	93813	86586	59854	01309
89215	46632	30988	79412	64601	22042	71379	05616
49880	60994	09374	10377	54878	80433	05994	58575
07468	27779	94664	39250	48561	54763	07733	73850
13742	93176	26563	62102	55681	16113	97148	24914
59149	94667	11891	63282	07489	04578	48465	82794

1. A survey of students in the eighth grade shows that 72% of them are wearing or have worn braces. Estimate the probability that 7 out of 10 eighth grade students wear braces or have worn braces. ____________

2. A driving school advertises that 88% of those taking their course pass their driving test. Estimate the probability that 9 out of 10 people who take the school's training will pass their driving test. ____________

3. An ice cream store stocks it shelves with 20% black raspberry chip ice cream because 20% of their customers choose that as their favorite flavor. Estimate the probability that 2 out of every 10 customers of the ice cream store will purchase black raspberry chip ice cream. ____________

4. In the first semester of the year, 61% of the class has been absent at least one day. Estimate the probability that 6 out of 10 students will be absent at least one day in the second semester. ____________

5. 42% of the students surveyed in the eighth grade have more than one sibling. Estimate the probability that 4 out of 10 students in the rest of the school have more than one sibling. ____________

Name ______________________ Date __________ Class __________

LESSON 10-3

Reteach

Use a Simulation

Situation: Strout's Market is having a contest. They give a puzzle piece to each customer at the checkout. A customer who collects all 10 different puzzle pieces gets $100 in store credit.

Using a table of random numbers, you can model the situation to estimate how many times a customer would have to shop to collect all 10 puzzle pieces.

- Start anywhere in the table. Count the numbers you pass as you "collect" the digits 0-9.

3	1	9	4	1	1	8	8
5	7	4	5	7	7	9	0
7	0	3	0	1	3	5	0
0	4	3	8	9	5	3	8
2	6	1	7	6	7	6	9
0	8	2	6	5	5	9	2

Suppose you start at the top of Column 3 and move to the right. List each number until you have collected all the numbers 0–9.

9 4 1 1 8 8 5 7 4 5 7 7 9 0 7 0 3 0 1 3 5 0 0 4 3 8 9 5 3 8 2 6

You had to go through 32 numbers to get each number at least once (underscored).

- Do the experiment again.

Suppose you start at the bottom of Column 4 and move to the right. When you reach the end of the row, go to the beginning of the table.

6 5 5 9 2 3 1 9 4 1 1 8 8 5 7 4 5 7 7 9 0

You had to go through 21 numbers to get each number at least once (underscored).

- Find the average of your results. $\frac{32 + 21}{2} = \frac{53}{2} = 26.5$

So, on average, you need to shop 27 times to get all 10 pieces to win $100 credit.

Model each situation. Use the list of random numbers shown above. Do two trials. Tell where you start for each trial.

1. A box of Whammos contains a toy dinosaur. If there are 10 different model dinosaurs in the collection, estimate how many boxes of Whammos you would have to buy to get all 10 dinosaurs.

____

____

2. For this spinner, estimate how many times you would have to spin the pointer to get the numbers 1–10.

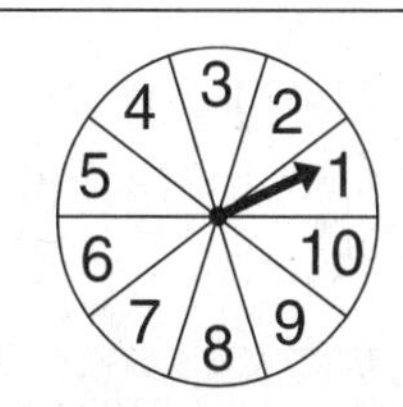

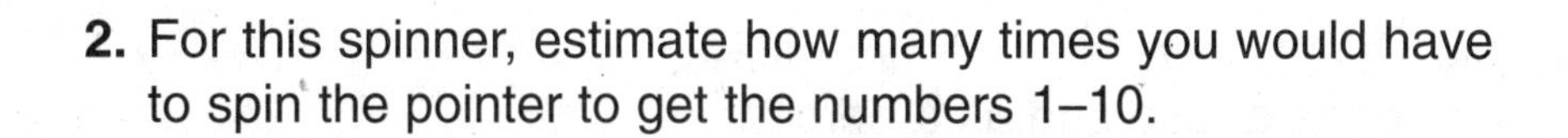

____

____

Name ______________________ Date ____________ Class ____________

LESSON 10-3

Challenge

Rolling and Tossing

To design a simulation, you may use different devices, such as number cubes or coins.

Situation: At Sonia's Spa, two-thirds of the female clients come to lose weight. For an article about spas, a female client at Sonia's was interviewed. What is the probability that this woman is at the spa to lose weight?

Simulation: To model a ratio of $\frac{2}{3}$, you can use a number cube so that 4 of the outcomes–1, 2, 3, 4–represent *came to lose weight and* 2 of the outcomes–5, 6–represent *did not come to lose weight.*

Then, P(came to lose weight) $= \frac{4}{6}$, or $\frac{2}{3}$.

To carry out this simulation, Kim rolled a number cube 10 times, with the following results: 4 5 2 6 6 6 1 2 4 3

1. How many of the 10 trials resulted in a woman who came to the spa to lose weight? ____________

2. Find P(came to lose weight). Answer as a ratio and as a percent. ____________

Situation: A study shows that a new medication has a 50% chance of curing the condition for which it is prescribed. Keith's doctor prescribes the medication for him. What is the probability that the medication will cure Keith's condition?

3. Using a cube numbered 1–6, describe a simulation.

____

____

4. Carry out your simulation for 10 trials. Calculate P(cures). Answer as a ratio and as a percent.

____

5. Using a coin, describe a simulation.

____

____

6. Carry out your simulation for 10 trials. Calculate P(cures). Answer as a ratio and as a percent.

____

Name ______________________ Date __________ Class __________

Problem Solving
Use a Simulation

Use the table of random numbers below. Use at least 10 trials to simulate each situation. Write the correct answer.

1. Of people 18–24 years of age, 49% do volunteer work. If 10 people ages 18–24 were chosen at random, estimate the probability that at least 4 of them do volunteer work.

____

87244	11632	85815	61766
19579	28186	18533	24633
74581	65633	54238	32848
87549	85976	13355	46498
53736	21616	86318	77291
24794	31119	48193	44869
86585	27919	65264	93557
94425	13325	16635	25840
18394	73266	67899	38783
94228	23426	76679	41256

2. In the 2000 Presidential election, 56% of the population of North Carolina voted for George W. Bush. If 10 people were chosen at random from North Carolina, estimate the probability that at least 8 of them voted for Bush.

____

3. Forty percent of households with televisions watched the 2001 Super Bowl game. If 10 households with televisions are chosen at random, estimate the probability that at least 3 watched the 2001 Super Bowl.

____

Use the table above and at least 10 trials to simulate each situation. Choose the letter for the best estimate.

4. As of August 2000, 42% of U.S. households had Internet access. If 10 households are chosen at random, estimate the probability that at least 5 of them will have Internet access.

   A 0% **C** 60%
   B 30% **D** 90%

5. On average, there is rain 20% of the days in April in Orlando, FL. Estimate the probability that it will rain at least once during your 7-day vacation in Orlando in April.

   F 20% **H** 70%
   G 50% **J** 40%

6. Kareem Abdul-Jabaar is the NBA lifetime leader in field goals. During his career, he made 56% of the field goals he attempted. In a given game, estimate the probability that he would make at least 6 out of 10 field goals.

   A 40% **C** 80%
   B 60% **D** 100%

7. At the University of Virginia 39% of the applicants are accepted. If 10 applicants to the University of Virginia are chosen at random, estimate the probability that at least 4 of them are accepted to the University of Virginia.

   F 10% **H** 80%
   G 40% **J** 70%

Name ______________________ Date ____________ Class ____________

LESSON 10-3

Reading Strategies

Analyze Information

A **simulation** is an experiment that models a real situation.

You can simulate flipping a coin without actually using a coin.

Random numbers can simulate flipping a coin. *Random* means happening by chance. No outcome is more likely than another.

You could set up an experiment using random numbers as follows:

- An even number shows an outcome of heads.
- An odd number shows an outcome of tails.

Here are five numbers that have occurred randomly:

9 6 9 1 6

1. Is 9 an even or odd number? What outcome does 9 represent?

____

2. Is 6 an even or odd number? What outcome does 6 represent?

____

3. Use H for heads and T for tails. List the outcomes for the five random numbers.

____

4. From the five outcomes, what is *P*(heads)? ________

5. From the five outcomes, what is the *P*(tails)? ________

This table of ten random numbers simulates flipping a coin. An even number represents heads. An odd number represents tails.

4 2 6 3 3 7 4 6 8 1

Use the random number table to answer the following questions.

8. Use H for heads and T for tails. List the outcomes in the table.

____

9. From these outcomes, what is *P*(heads)? ________

10. From these outcomes, what is *P*(tails)? ________

Name ______________________ Date ____________ Class ____________

LESSON 10-3

Puzzles, Twisters & Teasers

No Cheating!

Circle the words from the list that you find. Find a word that answers the riddle. Circle it and write it on the line.

simulation	random	numbers	model	set
table	digit	probability	reasonable	strategy

S I M U L A T I O N A R T
E T D I G I T E T U S E Y
T A R A N D O M P M D A U
G I T A B L E G H B F S I
N J I P T T G B U E G O P
F T C H E E T A H R H N O
B J I K O P G D E S J A L
M O D E L D U Y O P K B K
P R O B A B I L I T Y L V
Q W E R T Y U I O P H E B

What large animal cheats on a test? A ______________________.

Name ______________________ Date ____________ Class ____________

LESSON 10-4

Practice A

Theoretical Probability

An experiment consists of tossing two coins.

1. List all the possible outcomes. ____________

2. What is the probability of tossing a head and a tail? ____________

3. What is the probability of the outcomes being the same? ____________

An experiment consists of rolling a fair number cube. Find the probability of each event.

4. *P*(6) ____________

5. *P*(1) ____________

6. *P*(odd number) ____________

7. *P*(>4) ____________

Find the probability of each event using two number cubes.

8. *P*(rolling two 5s) ____________

9. *P*(total shown = 4) ____________

10. *P*(total shown = 2) ____________

11. *P*(total shown < 4) ____________

12. *P*(total shown > 11) ____________

13. *P*(rolling two even numbers) ____________

14. A bag contains 9 red marbles and 4 blue marbles. How many clear marbles should be added to the bag so the probability of drawing a red marble is $\frac{3}{5}$? ____________

15. In a game two fair number cubes are rolled. To make the first move, you need to roll an even total. What is the probability of rolling an even total? ____________

Name ______________________ Date __________ Class __________

LESSON 10-4

Practice B

Theoretical Probability

An experiment consists of rolling one fair number cube. Find the probability of each event.

1. $P(3)$ ____________

2. $P(7)$ ____________

3. $P(1 \text{ or } 4)$ ____________

4. $P(\text{not } 5)$ ____________

5. $P(< 5)$ ____________

6. $P(> 4)$ ____________

7. $P(2 \text{ or odd})$ ____________

8. $P(\leq 3)$ ____________

An experiment consists of rolling two fair number cubes. Find the probability of each event.

9. $P(\text{total shown} = 3)$ ____________

10. $P(\text{total shown} = 7)$ ____________

11. $P(\text{total shown} = 9)$ ____________

12. $P(\text{total shown} = 2)$ ____________

13. $P(\text{total shown} = 4)$ ____________

14. $P(\text{total shown} = 13)$ ____________

15. $P(\text{total shown} > 8)$ ____________

16. $P(\text{total shown} \leq 12)$ ____________

17. $P(\text{total shown} < 7)$ ____________

18. A bag contains 9 pennies, 8 nickels, and 5 dimes. How many quarters should be added to the bag so the probability of drawing a dime is $\frac{1}{6}$? ____________

19. In a game two fair number cubes are rolled. To make the first move, you need to roll a total of 6, 7, or 8. What is the probability that you will be able to make the first move? ____________

Name ______________________ Date ____________ Class ____________

LESSON 10-4

Practice C

Theoretical Probability

An experiment consists of rolling two fair number cubes. Find the probability of each event.

1. P(total shown = 5)

____

2. P(total shown > 3)

____

3. P(total shown > 10)

____

4. P(total shown < 12)

____

5. P(total shown ≥ 7)

____

6. P(total shown ≤ 4)

____

Three separate jars each contain 2 different color marbles. Jar A has a red and a blue marble. Jar B has a red and a green marble. Jar C has a purple and a white marble. One marble is drawn from each jar. The table shows a sample space with all outcomes equally likely. Find each probability.

Jar A	Jar B	Jar C	Outcome
R	R	P	RRP
R	R	W	RRW
R	G	P	RGP
R	G	W	RGW
B	R	P	BRP
B	R	W	BRW
B	G	P	BGP
B	G	W	BGW

7. P(RRP)

____

8. P(BGW)

____

9. P(2 red with another color)

____

10. P(a green with two other colors)

____

11. P(1 white or 1 purple)

____

12. A bag contains 12 red cubes, 15 blue cubes, 10 green cubes, and 14 yellow cubes. How many purple cubes should be added to the bag so the probability of drawing a blue cube is $\frac{1}{4}$? ______________

13. In a game two fair number cubes are rolled. To make the first move, you need to roll a total of 7, 8, or 9. What is the probability that you will be able to make the first move? ______________

Name ______________________ Date __________ Class __________

LESSON **10-4**

Reteach

Theoretical Probability

The sample space for a fair coin has 2 possible outcomes: heads or tails. Both possibilities have the same chance of occurring; they are **equally likely.**

The probability of each outcome is $\frac{1}{2}$.

$P(\text{heads}) = P(\text{tails}) = \frac{1}{2}$

Complete to find each probability.

1. 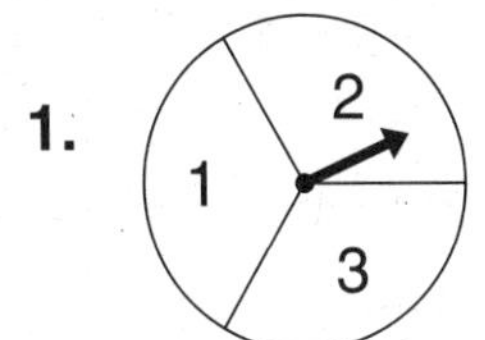

2. 

	1.	2.
How many outcomes in the sample space?	________	________
Are the outcomes equally likely?	________	________
What is the probability for each outcome?	________	________

For this spinner, there are 10 possible outcomes in the sample space. The outcomes are equally likely.

$P(7) = \frac{1}{10}$ $\quad P(\text{even number}) = \frac{5}{10}$, or $\frac{1}{2}$

$P(\text{a number greater than 4}) = \frac{6}{10}$, or $\frac{3}{5}$

When the possible outcomes are equally likely, you calculate the probability that an event E will occur by using a ratio.

$$\mathbf{P(E) = \frac{\text{number of favorable outcomes}}{\text{total number of possible outcomes}}}$$

Find each probability.

3. 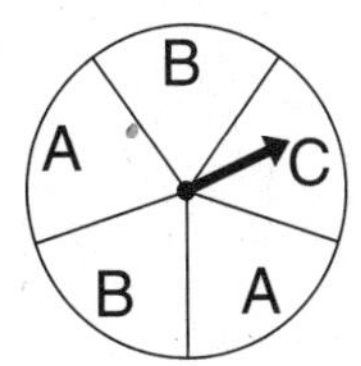

$P(C) =$ ____

$P(A) =$ ____

4. 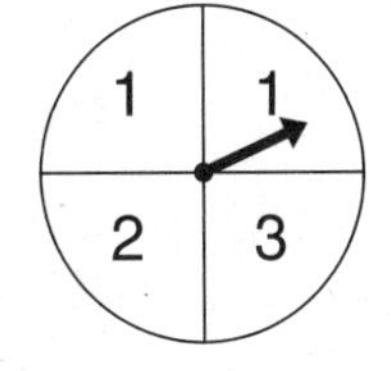

$P(1) =$ ____, or ____

$P(\text{even}) =$ ____

5. 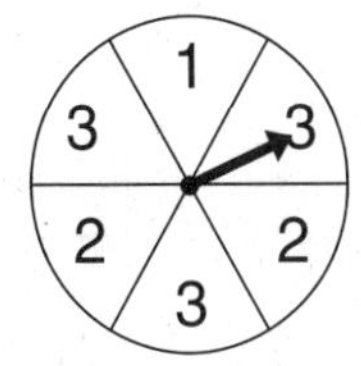

$P(2) =$ ____, or ____

$P(\text{odd}) =$ ____, or ____

Name ______________________ Date __________ Class __________

LESSON 10-4

Reteach

Theoretical Probability (continued)

For this spinner:

$P(\text{odd}) = \frac{3}{6}$, or $\frac{1}{2}$ $P(\text{even}) = \frac{1}{6}$

You cannot get an odd number and an even number in the same spin. $P(\text{odd } \textit{and} \text{ even}) = 0$

Events that cannot occur in the same trial are called **mutually exclusive.**

A number is drawn from {−6, −4, 0, 2, 4, 7, 9}. List the possible favorable results for each event. Tell if the events are mutually exclusive.

6. *Event A*: get an odd number

____

Event B: get a negative number

____

Are *A* and *B* mutually exclusive? Explain.

____

7. *Event C*: get a multiple of 3

____

Event D: get an even number

____

Are *C* and *D* mutually exclusive? Explain.

____

For the spinner at the top of this page:

$P(\text{odd}) = \frac{3}{6}$, or $\frac{1}{2}$ $P(\text{even}) = \frac{1}{6}$ $P(\text{odd } \textit{or} \text{ even}) = \frac{3}{6} + \frac{1}{6} = \frac{4}{6}$, or $\frac{2}{3}$

A number is drawn from {−6, −4, 0, 5, 6, 7, 9}. Find the indicated probabilities.

8. odd numbers are: 5, 7, 9

numbers < 0 are: ______________

$P(\text{odd}) = \frac{\quad}{7}$

$P(\text{number} < 0) = \frac{\quad}{7}$

$P(\text{odd number or number} < 0) =$

$\frac{\quad}{7} + \frac{\quad}{7} = \frac{\quad}{7}$

9. numbers > 6 are: ______________

even numbers: ______________

$P(\text{number} > 6) =$ ______________

$P(\text{even number}) =$ ______________

$P(\text{number} > 6 \text{ or even number}) =$

______ + ______ = ______

Name ______________________ Date ____________ Class ____________

LESSON **10-4**

Challenge
Picture This

Venn diagrams can be used to illustrate and solve problem situations involving probability.

A B
6 2 4 1 3

Consider a cube numbered 1–6.

Let *Event A* = rolling an even number on the cube.
favorable outcomes = 2, 4, 6

Let *Event B* = rolling a number less than 5 on the cube.
favorable outcomes = 1, 2, 3, 4

Note that the numbers 2 and 4 are in both events and, thus, lie in the intersection of the two circles that represent *Events A* and *B*.

So, to determine the probability of getting an even number that is also less than 5, the favorable outcomes are in the intersection of the circles.

$$P(A \text{ and } B) = \frac{\text{number of favorable outcomes}}{\text{total number of possible outcomes}} = \frac{2}{6}, \text{ or } \frac{1}{3}.$$

Then, to determine the probability of getting an even number *or* a number that is less than 5, count the elements in the intersection only once.

$$P(A \text{ or } B) = \frac{\text{number of favorable outcomes}}{\text{total number of possible outcomes}} = \frac{5}{6}$$

Draw a Venn diagram to solve each problem.
A cube numbered 1–6 is rolled once.

1. Find the probability of getting an odd number that is greater than 2.

   Event A = a number that is ______

   Event B = a number that is ______

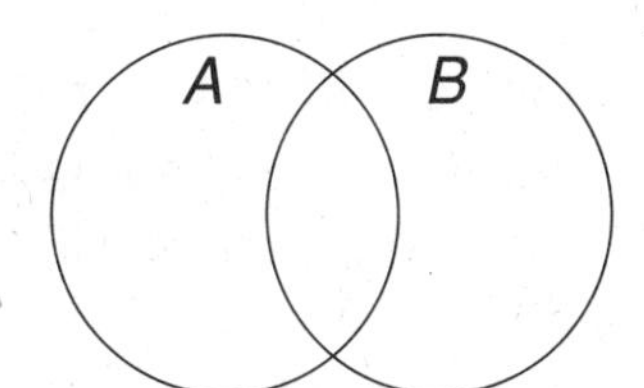

   $P(A \text{ and } B)$ = ______________

2. Find the probability of getting an even number or a number less than 3.

   Event A = a number that is ______

   Event B = a number that is ______

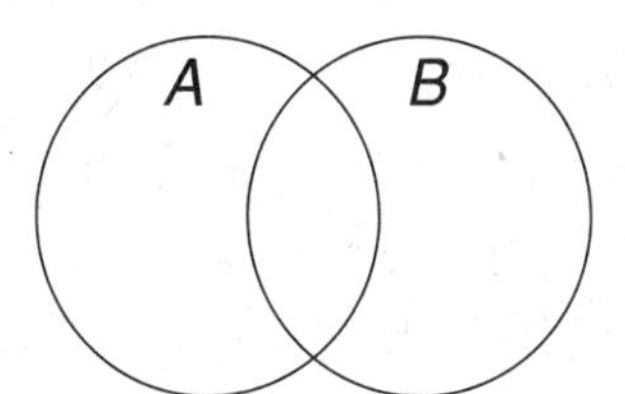

   $P(A \text{ or } B)$ = ______________

Name ______________________________ Date ___________ Class ___________

LESSON **10-4**

Problem Solving
Theoretical Probability

A company that sells frozen pizzas is running a promotional special. Out of the next 100,000 boxes of pizza produced, randomly chosen boxes will be prize winners. There will be one grand prize winner who will receive $100,000. Five hundred first prize winners will get $1000, and 3,000 second prize winners will get a free pizza. Write the correct answer in fraction and percent form.

1. What is the probability that the box of pizza you just bought will be a grand prize winner?

____

2. What is the probability that the box of pizza you just bought will be a first prize winner?

____

3. What is the probability that the box of pizza you just bought will be a second prize winner?

____

4. What is the probability that you will win anything with the box of pizza you just bought?

____

Researchers at the National Institutes of Health are recommending that instead of screening all people for certain diseases, they can use a Punnett square to identify the people who are most likely to have the disease. By only screening these people, the cost of screening will be less. Fill in the Punnett square below and use them to choose the letter for the best answer.

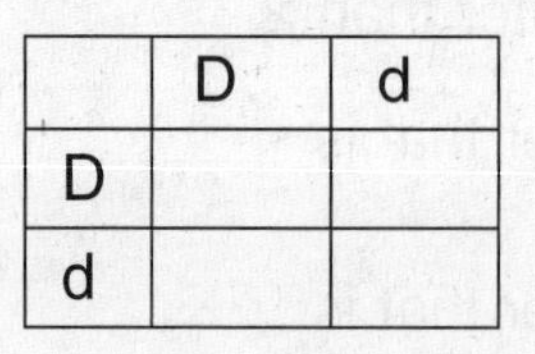

	D	d
D		
d		

5. What is the probability of DD?

   A 0% **C** 50%

   B 25% **D** 75%

6. What is the probability of Dd?

   F 25% **H** 75%

   G 50% **J** 100%

7. What is the probability of dd?

   A 0% **C** 50%

   B 25% **D** 75%

8. DD or Dd indicates that the patient will have the disease. What is the probability that the patient will have the disease?

   F 25% **H** 75%

   G 50% **J** 100%

Name ______________________ Date __________ Class __________

LESSON 10-4

Reading Strategies

Draw Conclusions

Theoretical probability describes what might be expected to happen in an event. It helps you draw conclusions.

$$\text{Probability(event)} = \frac{\text{number of ways an event can occur}}{\text{total number of events}}$$

When you flip a coin, there are two possible events.

→ land on heads

→ land on tails

You have a 1 out of 2 chance for each.

$P(\text{heads}) = \frac{1}{2}$ $\quad$ $P(\text{tails}) = \frac{1}{2}$

These events have the same probability, so you can draw the conclusion that they are **equally likely** to occur.

A number cube has six faces. The numbers on the six faces are 2, 3, 4, 2, 3, and 4. Answer the following about this number cube.

1. How many ways can the number cube land? __________

2. How many ways can you get a 2 on the cube? __________

3. What is P(rolling a 2)? __________

4. How many ways can you get an even number? __________

5. What is P(rolling an even number)? __________

6. Can you conclude that P(rolling a 2) and P(rolling an even number) are equally likely to occur? Explain.

____

____

7. What can you conclude about rolling an even number or an odd number?

____

Name ______________________ Date ____________ Class ____________

LESSON **10-4**

Puzzles, Twisters & Teasers

That's Odd!

Using the chart, write the probability for each outcome as a fraction. Unscramble the letters to answer the riddle. The fractions under the riddle will give you hints to get you started.

HTT Probability: ____ **E**

THT Probability: ____ **O**

TTT Probability: ____ **P**

HHT Probability: ____ **F**

HHH Probability: ____ **A**

TTH Probability: ____ **S**

HTH Probability: ____ **R**

THH Probability: ____ **K**

TH Probability: ____ **C**

THHT Probability: ____ **Y**

Penny	Dime	Quarter	Outcome
H	H	H	HHH
H	H	T	HHT
H	T	T	HTT
H	T	H	HTH
T	H	H	THH
T	H	T	THT
T	T	H	TTH
T	T	T	TTT

What did the boy do when he wanted to see time fly?

He dropped his clock

____ ____ **F** **A**

$\frac{1}{8}$ $\frac{1}{4}$

____ ____ ____ **S** ____ ____ ____ ____ ____ **R**

$\frac{1}{4}$ $\frac{1}{4}$ 0 0 $\frac{1}{8}$ $\frac{1}{8}$ $\frac{1}{8}$ $\frac{1}{4}$

Name ______________________ Date __________ Class __________

Practice A
Independent and Dependent Events

Determine if the events are dependent or independent.

1. drawing a card from a deck of cards and tossing a coin

____

2. drawing two cards from a regular deck of cards and not replacing the first

____

3. spinning a number on a spinner and drawing a marble from a container

____

4. drawing two red marbles without replacement from a container of red and blue marbles

____

An experiment consists of spinning each spinner once. Find the probability. For each spin, all outcomes are equally likely.

5. *P*(A and 2) _______________

6. *P*(D and 1) _______________

7. *P*(B and 3) _______________

8. *P*(B and 1 or 3) _______________

9. *P*(C and 1 or 2) _______________

10. *P*(A and not 1) _______________

2 1 2 1 3 1

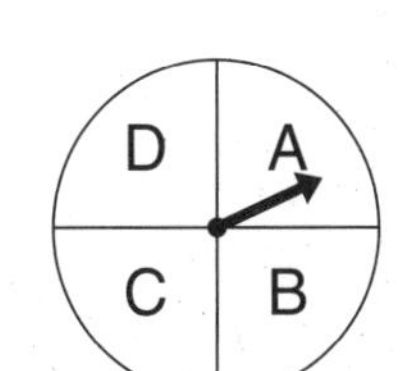

11. Georgiana wants to toss three coins and get all heads. What is the probability of tossing 3 coins and getting 3 heads?

____

Name ______________________ Date __________ Class __________

LESSON 10-5

Practice B

Independent and Dependent Events

Determine if the events are dependent or independent.

1. choosing a tie and shirt from the closet ______________

2. choosing a month and tossing a coin ______________

3. rolling two fair number cubes once, then rolling them again if you received the same number on both number cubes on the first roll ______________

An experiment consists of rolling a fair number cube and tossing a fair coin.

4. Find the probability of getting a 5 on the number cube and tails on the dime. ________

5. Find the probability of getting an even number on the number cube and heads on the dime. ________

6. Find the probability of getting a 2 or 3 on the number cube and heads on the dime. ________

A box contains 3 red marbles, 6 blue marbles, and 1 white marble. The marbles are selected at random, one at a time, and are not replaced. Find the probability.

7. *P*(blue and red) ______________

8. *P*(white and blue) ______________

9. *P*(red and white) ______________

10. *P*(red and white and blue) ______________

11. *P*(red and red and blue) ______________

12. *P*(red and blue and blue) ______________

13. *P*(red and red and red) ______________

14. *P*(white and blue and blue) ______________

15. *P*(white and red and white) ______________

Name ______________________ Date __________ Class __________

LESSON 10-5

Practice C

Independent and Dependent Events

Consider a regular deck of cards without the jokers. Cards are replaced after each draw. Find the probability of each of the following.

1. *P*(pair of red kings)

____

2. *P*(a diamond and a black seven)

____

Use the same deck of cards but do not replace the card after each draw.

3. *P*(ace of hearts and king of hearts)

____

4. *P*(a ten and a jack)

____

5. *P*(red card and a black card)

____

6. *P*(a club and king or red ace)

____

7. Mr. and Mrs. Reginald are expecting their first baby. The doctor tells them they are having triplets. What is the probability that the babies will all be the same sex?

____

8. Sid has a bag of 12 red, 14 brown, and 10 blue marbles. He chooses one, shoots it, and chooses another. What is the probability that his first selection is a red marble, and then a blue marble?

____

9. If Justine's initials are JMD, what is the probability that she will draw her initials from a box containing the letters of the alphabet? There is no replacement of letters after each is drawn.

____

10. There are 13 math students, 10 science students, and 17 English students in a group. If only one prize is allowed per person, what is the probability that the moderator will award a science student a prize and then award another prize to a math student?

____

Name ______________________ Date ____________ Class ____________

LESSON 10-5

Reteach

Independent and Dependent Events

Carlos is to draw 2 straws at random from a box of straws that contains 4 red, 4 white, and 4 striped straws.

$P(\text{1st straw is striped}) = \frac{4}{12}$ ← number of striped straws / ← total number of straws

If Carlos *returns* the 1st straw to the box before drawing the 2nd straw, the probability that the 2nd straw is striped remains the same.

$P(\text{2nd straw is striped})$

$= \frac{4}{12}$ ← same number of striped straws / ← same total number of straws

When the 1st straw is returned before the 2nd draw, the 2nd draw occurs as though the 1st draw never happened, **independent events.**

$P(\text{striped and striped}) = \frac{4}{12} \times \frac{4}{12}$

$= \frac{1}{3} \times \frac{1}{3} = \frac{1}{9}$

If Carlos *does not return* the first straw to the box before drawing the second straw, the probability that the second straw is striped changes.

$P(\text{2nd straw is striped})$

$= \frac{3}{11}$ ← one striped straw has been taken / ← one less straw in total number

When the 1st straw is not returned before the 2nd draw, the number of straws remaining is changed, **dependent events.**

$P(\text{striped and striped}) = \frac{4}{12} \times \frac{3}{11}$

$= \frac{1}{3} \times \frac{3}{11} = \frac{1}{11}$

Describe the events as independent or dependent.

1. Josh tosses a coin and spins a spinner. ____________

2. Ana draws a colored toothpick from a jar. Without replacing it, she draws a second toothpick. ____________

3. Sue draws a card from a deck of cards and replaces it. Then she draws a second card from the deck. ____________

Each situation begins with a box of marbles that contains 2 red, 3 blue, 4 green, and 3 yellow marbles. Complete to find each probability.

4. A 1st marble is drawn and replaced. Then a 2nd marble is drawn.

   $P(\text{red and blue}) = \frac{\_\_}{12} \times \frac{\_\_}{12} =$ ______

5. A 1st marble is drawn and not replaced. A 2nd marble is drawn.

   $P(\text{red and blue}) = \frac{\_\_}{12} \times \frac{\_\_}{11} =$ ______

6. A 1st marble is drawn and replaced. Then a 2nd marble is drawn.

   $P(\text{red and red}) = \frac{\_\_}{12} \times \frac{\_\_}{\_\_} =$ ______

7. A 1st marble is drawn and not replaced. A 2nd marble is drawn.

   $P(\text{red and red}) = \frac{\_\_}{12} \times \frac{\_\_}{\_\_} =$ ______

Name ______________________ Date __________ Class __________

LESSON 10-5 Challenge
Probability from a Table

The table shows the results of a survey of 50 students. The students were asked whether they liked a newly released movie.

	Yes	No
Male	16	14
Female	12	8

According to the table:

16 + 14, or 30 males were surveyed.
12 + 8, or 20 females were surveyed.
16 + 12, or 28, of those surveyed liked the movie.
14 + 8, or 22, of those surveyed did not like the movie.

The probability that a student randomly selected from the group is a male is $\frac{30}{50}$, or $\frac{3}{5}$.

The probability that a student did not like the movie is $\frac{22}{50}$, or $\frac{11}{25}$.

The probability that a student is a male who did not like the movie is $\frac{14}{50}$, or $\frac{7}{25}$.

The table below shows the results of a survey of 50 students. The students were asked whether they liked a newly released CD. One student was selected at random from this group. Use the table to solve.

	Yes	No
Male	35	5
Female	8	2

1. What is the probability that the selected student liked the CD? ________

2. What is the probability that the student did not like the CD? ________

3. What is the probability that the student is a female? ________

4. What is the probability that the student is a female who liked the movie? ________

5. What is the probability that the student is a male who did not like the movie? ________

Name ______________________ Date __________ Class __________

LESSON 10-5

Problem Solving

Independent and Dependent Events

Are the events independent or dependent? Write the correct answer.

1. Selecting a piece of fruit, then choosing a drink.

____

2. Buying a CD, then going to another store to buy a video tape if you have enough money left.

____

Dr. Fred Hoppe of McMaster University claims that the probability of winning a pick 6 number game where six numbers are drawn from the set 1 through 49 is about the same as getting 24 heads in a row when you flip a fair coin.

3. Find the probability of winning the pick 6 game and the probability of getting 24 heads in a row when you flip a fair coin.

____

____

4. Which is more likely: to win a pick 6 game or to get 24 heads in a row when you flip a fair coin?

____

____

In a shipment of 20 computers, 3 are defective. Choose the letter for the best answer.

5. Three computers are randomly selected and tested. What is the probability that all three are defective if the first and second ones are not replaced after being tested?

A $\frac{1}{760}$ C $\frac{27}{8000}$

B $\frac{1}{1140}$ D $\frac{3}{5000}$

6. Three computers are randomly selected and tested. What is the probability that all three are defective if the first and second ones are replaced after being tested?

F $\frac{1}{760}$ H $\frac{27}{8000}$

G $\frac{1}{1140}$ J $\frac{3}{5000}$

7. Three computers are randomly selected and tested. What is the probability that none are defective if the first and second ones are not replaced after being tested?

A $\frac{34}{57}$ C $\frac{4913}{6840}$

B $\frac{4913}{8000}$ D $\frac{1}{2000}$

8. Three computers are randomly selected and tested. What is the probability that none are defective if the first and second ones are replaced after being tested?

F $\frac{34}{57}$ H $\frac{4913}{6840}$

G $\frac{4913}{8000}$ J $\frac{1}{2000}$

Name ______________________ Date ____________ Class ____________

LESSON 10-5

Reading Strategies

Use Context

When the outcome of one event does not affect the outcome of another, they are called **independent events.**

A cube has the numbers 1–6 on the faces.

P(rolling a 6) is $\frac{1}{6}$.

If you roll the cube a second time, P(rolling a 6) is still $\frac{1}{6}$.

The second roll of the cube is independent of the first roll.

When the outcome of one event affects the probability of another, they are called **dependent events.**

There are 10 cubes in a bag. Three of them are red.

P(drawing a red cube) is $\frac{3}{10}$.

If a red cube is drawn and not replaced, the probability of drawing a red cube on the second draw is 2 out of 9.

So P(drawing another red cube) is $\frac{2}{9}$.

Drawing a cube the second time is dependent on drawing the first cube.

Tell if each situation describes dependent events or independent events.

1. You flip a coin. Then you flip a coin a second time.

____

2. You roll a number cube three times.

____

3. You draw a card from a deck of cards and do not replace the card. Then you draw a second time.

____

4. You spin a spinner once. Then you spin the spinner a second time.

____

5. You pull a marble out of a jar and leave it on the table. Then you pull another marble out of the jar.

____

Name ______________________ Date ____________ Class ____________

LESSON 10-5

Puzzles, Twisters & Teasers

Declare Your Independence!

Decide whether the events are dependent or independent. Circle the letter above your answer. Unscramble the letters to answer the riddle.

1. tossing a coin twice
   Q dependent — I independent
2. pulling two socks from a drawer at the same time
   E dependent — P independent
3. drawing two marbles out of a bag
   B dependent — D independent
4. spinning a spinner five times
   M dependent — C independent
5. spinning two different spinners two times
   K dependent — E independent
6. drawing names out of a hat
   R dependent — B independent
7. throwing a pair of number cubes ten times
   X dependent — U independent
8. throwing one number cube five times
   U dependent — S independent
9. picking cards from a deck
   G dependent — V independent
10. throwing three coins three times
    A dependent — R independent

What do penguins eat for lunch?

___ ___ ___ ___ ___ ___ ___ ___ ___ ___

Name ______________________ Date __________ Class __________

LESSON **10-6**

Practice A

Making Decisions and Predictions

The zoo store sells caps with different animals pictured on the cap. The table shows the animals pictured on the last 100 caps sold. The manager plans to order 1500 new caps.

Animal Caps Sold

Animal	Number
Tiger	30
Orangutan	20
Panda Bear	25
Giraffe	18
Gazelle	7

1. Find the probability of selling a tiger cap.

____

2. How many tiger caps should the manager order?

____

3. Find the probability of selling a panda bear cap.

____

4. How many panda bear caps should the manager order?

____

5. Use probability to decide how many orangutan caps the manager should order.

____

6. Nancy spins the spinner at the right 60 times. Predict how many times the spinner will land on the number 2.

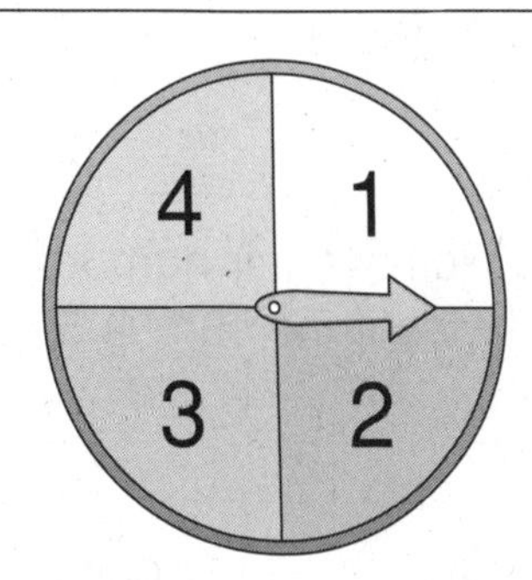

____

Decide whether the game is fair.

7. Roll two fair number cubes labeled 1–6. Player A wins if both numbers are odd.
Player B wins if both numbers are even.

____

Name ______________________ Date __________ Class __________

LESSON **10-6**

Practice B

Making Decisions and Predictions

A sports store sells water bottles in different colors. The table shows the colors of the last 200 water bottles sold. The manager plans to order 1800 new water bottles.

Water Bottles Sold

Color	Number
Red	30
Blue	50
Green	25
Yellow	10
Purple	10
Clear	75

1. How many red water bottles should the manager order? __________

2. How many green water bottles should the manager order? __________

3. How many clear water bottles should the manager order? __________

4. If the carnival spinner lands on 10, the player gets a large stuffed animal. Suppose the spinner is spun 30 times. Predict how many large stuffed animals will be given away. __________

Decide whether the game is fair.

5. Roll two fair number cubes labeled 1–6. Player A wins if both numbers are the same. Player B wins if both numbers are different.

____

6. Roll two fair number cubes labeled 1–6. Add the numbers. Player A wins if the sum is 5 or less. Player B wins if the sum is 9 or more.

____

7. Toss three fair coins. Player A wins if exactly one tail lands up. Otherwise, Player B wins.

____

Name ______________________ Date __________ Class __________

LESSON **10-6**

Practice C

Making Decisions and Predictions

A fair number cube is labeled 1–6. Predict the number of outcomes for the given number of rolls.

1. outcome: 5
number of rolls: 42

____

2. outcome: less than 3
number of rolls: 75

____

3. outcome: not 4
number of rolls: 60

____

4. outcome: 2, 3, or 4
number of rolls: 30

____

5. outcome: greater than 2
number of rolls: 48

____

6. outcome: multiple of 3
number of rolls: 90

____

7. In his last eight 5K runs, Jeremy had the following times in minutes: 24:48, 23:45, 23:12, 24:08, 25:36, 22:03, 23:29, and 24:01. Based on these results, what is the best prediction of the number of times Jeremy will run faster than 24 minutes in his next 20 5K runs?

____

8. Before a school vote on a mascot for a community river project, a sample of students surveyed gave the otter 18 votes, the osprey 8 votes, and the beaver 6 votes. Based on these results, predict the number of votes for each animal if 1200 students vote.

____

Decide whether each game is fair.

9. A spinner is divided evenly into 6 sections. There are 3 green sections, 2 blue sections, and 1 yellow section. Player A wins if the spinner does not land on green. Otherwise, Player B wins.

____

10. Roll two fair number cubes labeled 1–6. Add the numbers. Player A wins if the sum is 8 or more. Player B wins if the sum is 5 or less.

____

11. Toss three fair coins. Player A wins if exactly two tails land up. Player B wins if all heads or all tails land up.

____

Name ______________________ Date __________ Class __________

LESSON **10-6**

Reteach

Making Decisions and Predictions

Probability can be used to make predictions about data.

The spinner has 5 equal sections. To predict how many times you will land on the number 1 in 30 spins, first find the probability of landing on 1 in one spin. Then multiply 30 spins by that probability.

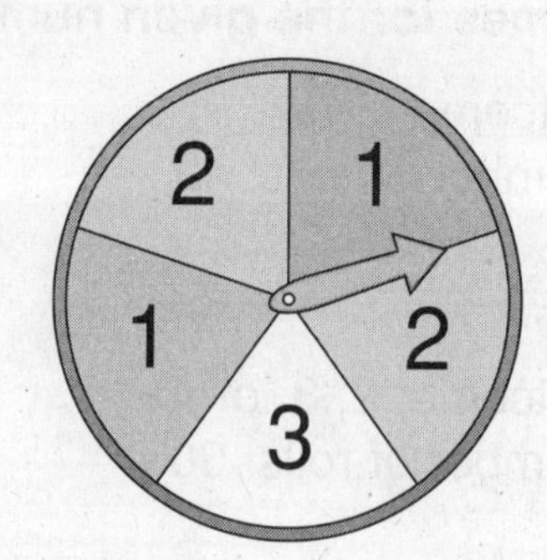

Step 1 Find the probability.

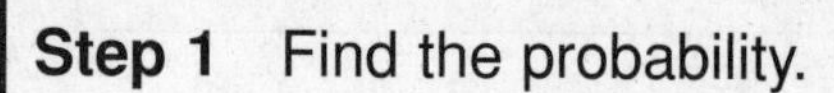

$$P(1) = \frac{\text{number of 1s}}{\text{number of equal sections}} = \frac{2}{5}$$

Step 2 Multiply the number of spins by the probability.

There are 30 spins and $P(1) = \frac{2}{5}$.

$30 \times \frac{2}{5} = 6 \times 2 = 12$

You will spin about 12 1s in 30 spins.

Each situation begins with a box of marbles that contains 2 red, 3 blue, 4 green, and 3 yellow marbles. Complete to find each probability.

1. Predict how many times you will land on C in 32 spins.

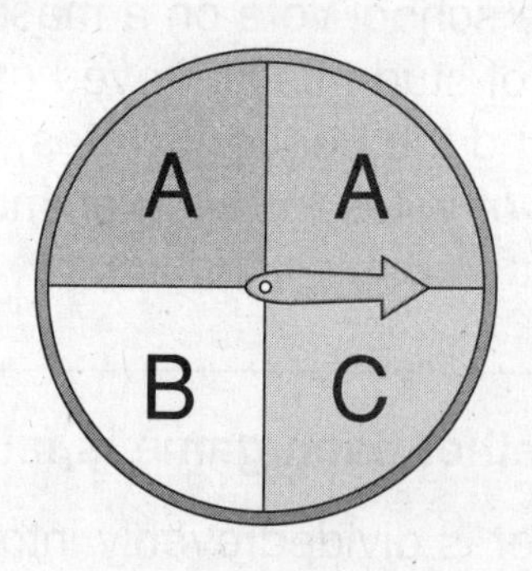

a. $P(C)$ = __________

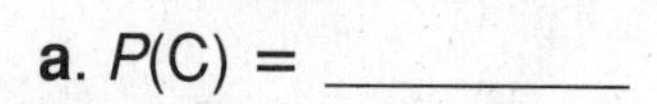

b. $32 \times P(C) = 32 \times$ __________ = __________

c. about __________ times

2. Predict how many times you will land on B in 48 spins.

a. $P(B)$ = __________

b. $48 \times P(B)$ = __________ = __________

c. about __________ times

3. Predict how many times you will land on A in 50 spins.

a. $P(A)$ = __________

b. __________ = __________

c. about __________ times

Name ______________________ Date __________ Class __________

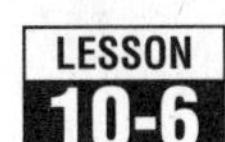

Challenge

Complimentary Events

The complement of an event A is the event that A does not occur. It includes all outcomes not in A. Represent the complement of *A* as not *A*.

Since either *A* or not *A* must occur, $P(A) + P(\text{not } A) = 1$. This means that $P(A) = 1 - P(\text{not } A)$ and $P(\text{not } A) = 1 - P(A)$.

Suppose *A* is the event rolling 4 when rolling a fair number cube. Then the complement of *A* is the event rolling a number that is not 4.

$$P(4) = \frac{1}{6} \qquad P(\text{not } 4) = 1 - \frac{1}{6} = \frac{5}{6}$$

Two fair number cubes are rolled. Solve.

1. Suppose *A* is the event rolling a sum of 2 or 3.

a. Find $P(A)$ = __________

b. Use words to describe the complement of *A*.

____

c. Find $P(\text{not } A)$ __________

2. Suppose *A* is the event rolling a sum of at least 10.

a. $P(A)$ = __________

b. Use words to describe the complement of *A*.

____

c. Find $P(\text{not } A)$. __________

3. Suppose *A* is the event rolling a sum of 8.

a. $P(A)$ = __________

b. Use words to describe the complement of *A*.

____

c. Find $P(\text{not } A)$. __________

Name ______________________ Date __________ Class __________

LESSON 10-6

Problem Solving

Making Decisions and Predictions

Write the correct answer.

1. A quality control inspector at a light bulb factory finds 2 defective bulbs in a batch of 1000 light bulbs. If the plant manufactures 75,000 light bulbs in one day, predict how many will be defective.

____

2. A game consists of rolling two fair number cubes labeled 1–6. Add both numbers. Player A wins if the sum is greater than 10. Player B wins if the sum is 7. Is the game fair or not? Explain.

____

3. A spinner has 5 equal sections numbered 1–5. Predict how many times Kevin will spin an even number in 40 spins.

____

4. In her last six 100-meter runs, Lee had the following times in seconds: 12:04, 13:11, 12:25, 11:58, 12:37, and 13:20. Based on these results, what is the best prediction of the number of times Lee will run faster than 13 seconds in her next 30 runs?

____

Use the table below that shows the number of colors of the last 200 T-shirts sold at a T-shirt shop. The manager of the store wants to order 1800 new T-shirts. Choose the letter of the best answer

T-Shirts Sold

Color	Number
Red	35
Blue	55
Green	15
Black	65
White	30

5. How many red T-shirts should the manager order?

A 175 **C** 378

B 315 **D** 630

6. How many blue T-shirts should the manager order?

F 495 **H** 900

G 665 **J** 990

7. How many more black T-shirts than white T-shirts should the manager order?

A 855 **C** 315

B 585 **D** 270

Name ______________________ Date __________ Class __________

Reading Strategies
Draw Conclusions

LESSON 10-6

You can use probability to make decisions about data.

The table shows the colors of 100 wrist bands sold at a sports shop.

Wrist Bands Sold

Color	Number
Yellow	40
Blue	20
Pink	30
Red	10

The store manager wants to order 2000 new bands. She can use the data in the table to decide how many of each color of wrist band to order in the next order. The manager should order the greatest number of yellow bands and the fewest number of red bands.

To find the number of red bands, first use the data in the table to find *P*(*red*), the probability of selling a red band.

$$P(red) = \frac{10}{100} = 0.1$$

Then use the probability to find how many red bands to order:

$$0.1 \text{ of } 2000 = 0.1 \times 2000 = 200$$

So, the manager should order 200 red bands.

Answer each question. Use the information above.

1. What is the probability of selling a yellow band?

____

2. Write an expression that represents how many yellow bands the manager should order.

____

3. How many yellow bands should the manager order?

____

4. What is the probability of selling a pink band?

____

5. How many pink bands should the manager order?

____

Name ______________________ Date __________ Class __________

LESSON 10-6

Puzzles, Twisters & Teasers

A Friendly Solution!

Find and circle words from the list in the word search horizontally, vertically, or diagonally. Find a word that answers the riddle. Circle it and write it on the line.

probability	decision	prediction	fair	unfair
game	proportion	chance	likely	favorable

```
H G R E C O M D A R E I G N P
A A D I T C V E I G V E R S R
F V G U E H T C S T A B E S E
A E L N V A T I L O S M A E D
V F R I E N D S H I P A E L I
O S A R L C O I P O K D E R C
R N O I P E S O T E L E R V T
A P C H R S U N F A I R L T I
B R P R O P O R T I O N A Y O
L O V P R O B A B I L I T Y N
E C H A F A V N C E B I L U N
```

What kind of ship never sinks? ______________ ________

Name ______________________ Date ____________ Class ____________

Practice A
Odds

1. Complete the table by finding the odds in favor and the odds against an event based on the probability of the event.

Probability	$\frac{1}{4}$	$\frac{2}{5}$	$\frac{1}{8}$	$\frac{3}{4}$	$\frac{3}{7}$
Odds in favor					
Odds against					

2. Complete the table with the missing information.

Probability	$\frac{3}{8}$				
Odds in favor		2:1	1:7	8:1	3:7
Odds against		1:2	7:1	1:8	7:3

If there are 28 boys and 22 girls in the music class, find the odds for each of the following.

3. What are the odds in favor of selecting a boy as the conductor?

____

4. What are the odds in favor of selecting a girl as the conductor?

____

5. What are the odds against selecting a boy as the conductor?

____

6. What are the odds against selecting a girl as the conductor?

____

7. If the probability of drawing a red card from a regular deck of cards without the jokers is $\frac{1}{2}$ what are the odds in favor of and against drawing a red card?

____

Name ______________________ Date ____________ Class ____________

LESSON 10-7

Practice B

Odds

A bag contains 9 red marbles, 5 green marbles, and 6 purple marbles.

1. Find P(red marble)

2. Find P(green marble)

3. Find P(purple marble)

4. Find the odds in favor of choosing a red marble.

5. Find the odds against choosing a red marble.

6. Find the odds in favor of choosing a green marble.

7. Find the odds against choosing a green marble.

8. Find the odds in favor of choosing a purple marble.

9. Find the odds against choosing a purple marble.

10. Find the odds in favor of not choosing a green marble.

11. Find the odds in favor of choosing a red or purple marble.

12. If the probability of Helena winning the contest is $\frac{2}{5}$, what are the odds in favor of Helena winning the contest?

13. The odds in favor of the Bruins winning the Stanley Cup are 5 to 4. What is the probability that the Bruins will win the Stanley Cup?

Name ________________________ Date ____________ Class ____________

LESSON 10-7

Practice C

Odds

Use the spinner to find the following odds. The spinner turns but the pointer stays in one place.

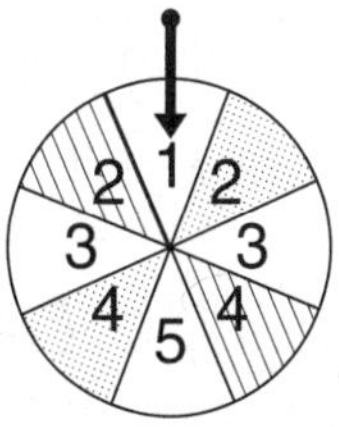

1. Find the odds in favor of the spinner stopping at 1.

____

2. Find the odds against the spinner stopping at 5.

____

3. Find the odds in favor of the spinner stopping at 2.

____

4. Find the odds against the spinner stopping at white.

____

5. Find the odds in favor of the spinner stopping at dots or 1.

____

6. Find the odds of the spinner stopping at lines and an even number.

____

Katera won a contest in math class. As her prize she could pick one envelope from 18 different envelopes. The prizes included a pass to the local amusement park, 6 movie passes, a gift certificate for a school hat, 8 free lunch passes, and 2 gift certificates to the school supply store. Each envelope contained one prize.

Find the odds in favor of each of the following.

7. Katera choosing the envelope containing the amusement park pass.

____

8. Katera choosing an envelope containing a free lunch pass.

____

9. Katera choosing an envelope containing a gift certificate for the school supply store.

____

10. Katera choosing an envelope containing a movie pass.

____

Name ______________________ Date __________ Class __________

LESSON 10-7

Reteach

Odds

Baseball fans do not usually ask "What is the probability that the New York Yankees will win the World Series this year?"

Fans who want to know the chances of a team winning usually ask "What are the *odds* that the Yankees will win?"

Odds that an event E will or will not occur can be defined as a ratio of probabilities.

$$\textbf{odds in favor} = \frac{P(E)}{P(\text{not } E)} \qquad \textbf{odds against} = \frac{P(\text{not } E)}{P(E)}$$

What are the odds in favor of getting a 4 in one roll of a numbered cube?

$$P(4) = \frac{1}{6} \quad P(\text{not } 4) = \frac{5}{6} \quad \text{odds}(4) = \frac{P(4)}{P(\text{not } 4)} = \frac{\frac{1}{6}}{\frac{5}{6}} = \frac{1}{5}$$

So, the odds in favor of getting a 4 are 1 to 5.

Complete to find the indicated odds. In each case, a cube numbered 1–6 is rolled once.

1. Find the odds in favor of getting a number greater than 4.

$P(> 4) =$ ____ $P(\text{not} > 4) =$ ____ $\text{odds}(> 4) = \frac{P(>4)}{P(\text{not} > 4)} =$ ____ $=$ ____, or ____

So, the odds in favor of getting a number greater than 4 are ____ to ____.

2. Find the odds against getting a 3.

$P(3) =$ ____ $P(\text{not } 3) =$ ____ $\text{odds}(\text{not } 3) = \frac{P(\text{not } 3)}{P(3)} =$ ____ $=$ ____

So, the odds against getting a 3 are ____ to ____.

3. Find the odds in favor of getting an even number.

$P(\text{even}) =$ ____ $P(\text{not even}) =$ ____ $\text{odds}(\text{even}) = \frac{P(\text{even})}{P(\text{not even})} =$ ____ $=$ ____, or ____

So, the odds in favor of getting an even number are ____ to ____.

Name ______________________________ Date ____________ Class ____________

LESSON **10-7**

Reteach

Odds, continued

Convert each probability ratio to an odds ratio.

4. The probability of winning a prize is $P(\text{win}) = \frac{2}{25}$.

$P(\text{not win}) = 1 - \frac{2}{25}$, or $\frac{\quad}{25}$ odds(win) $= \frac{P(\text{win})}{P(\text{not win})} = \frac{\frac{2}{25}}{\frac{\quad}{25}}$ or $\frac{2}{\quad}$, or 2 to ____

5. The probability of not winning a prize $P(\text{not win}) = \frac{100}{109}$.

$P(\text{win}) = 1 - \frac{100}{109}$, or $\frac{\quad}{109}$ odds(not win) $= \frac{P(\text{not win})}{P(\text{win})} = \frac{\frac{100}{109}}{\frac{\quad}{109}}$ or ____, or 100 to ____

Reconsider a problem in finding odds by using probabilities.

What are the odds in favor of getting a 4 in one roll of a numbered cube?

$P(4) = \frac{1}{6}$ $P(\text{not } 4) = \frac{5}{6}$ odds(4) $= \frac{P(4)}{P(\text{not } 4)} = \frac{\frac{1}{6}}{\frac{5}{6}} = \frac{1}{5}$

Study the ratio for odds with respect to its relation to the probability ratios.

$\frac{1}{5}$ ← numerator of probability in favor $1 + 5 = 6$ ← denominator of probability ratio

If the odds in favor of an event are $\frac{a}{b}$, or $a{:}b$, then the probability that the event will occur is $\frac{a}{a+b}$.

If the odds in favor of being chosen for a committee are 2:3, then the probability of being chosen is $\frac{2}{2+3}$, or $\frac{2}{5}$.

Convert each odds ratio to a probability ratio.

6. The odds in favor of winning a prize are 3:20.

The probability of winning a prize is: $\frac{3}{3 + \quad}$, or ____.

7. The odds against winning a prize are 100:7.

The probability of not winning a prize is: $\frac{100}{\quad}$, or ____.

Name ______________________ Date ____________ Class ____________

LESSON 10-7

Challenge

All Sizes and Shapes

The most common shaped die is a cube numbered 1 through 6. However, dice come in a variety of shapes. The illustrations show a cube and five other polyhedral dice. All but one of these six dice are regular polyhedrons.

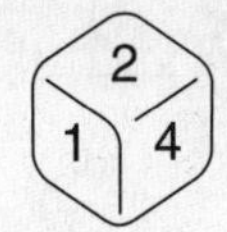

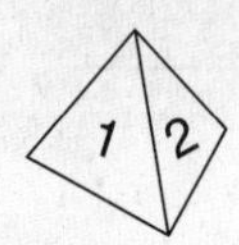

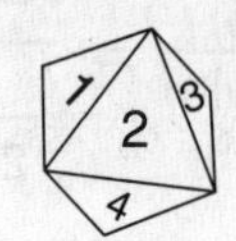

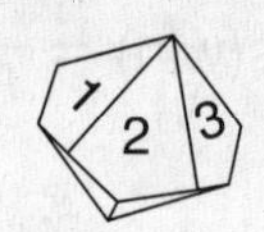

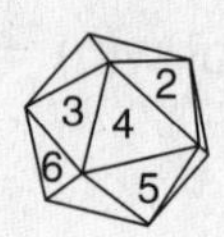

The table below shows the probability of rolling a 1 and the odds in favor of rolling a 1, not a 1, an even number, and a number that is a multiple of 5.

	A	B	C	D	E	F	G
1	Number of sides	Name of shape	*P*(1)	Odds(1)	Odds (not 1)	Odds (even)	Odds (multiple of 5)
2	4	tetrahedron	0.25				
3	6	cube		1:5			
4	8	octahedron			7:1		
5	10	decahedron				1:1	
6	12	dodecahedron					1:5

Answer each question.

1. The formula for cell C2 is $\frac{1}{A2}$. What does the A2 represent in the probability formula?

____

2. The first number in the ratio in cell F5 is $\frac{A5}{2}$. What does this number represent?

____

3. Complete the rest of the table.

Name ______________________ Date __________ Class __________

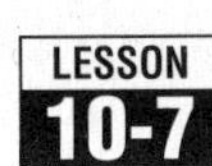

LESSON 10-7

Problem Solving

Odds

In the last 25 Summer Olympics since 1900, an American man has won the gold medal in the 400-meter dash 18 times. Write the correct answer.

1. Find the probability that an American man will win the gold medal in the 400-meter dash in the next Summer Olympics.

____

2. Find the probability that an American man will not win the gold medal in the 400-meter dash in the next Summer Olympics.

____

3. Find the odds that an American man will win the gold medal in the 400-meter dash in the next Summer Olympics.

____

4. Find the odds that an American man will not win the gold medal in the 400-meter dash in the next Summer Olympics.

____

Use the table below that shows the probability that a player will end up on a certain square after a single roll in a game of Monopoly.

Probability of Ending Up on a Monopoly Square

Square	Probability	Rank
In Jail	$\frac{39}{1000}$	1
Illinois Ave.	$\frac{32}{1000}$	2
Go	$\frac{31}{1000}$	3
Boardwalk	$\frac{26}{1000}$	18
Park Place	$\frac{22}{1000}$	33

5. What are the odds that you will end up in jail on your next roll in a game of Monopoly?

A 39:1000
B 39:961
C 1000:961
D 961:39

6. What are the odds that you will end up on Boardwalk on your next roll in a game of Monopoly?

A 13:500 **C** 13:487
B 500:13 **D** 487:13

7. What are the odds that you will not end up on Boardwalk on your next roll in a game of Monopoly?

F 487:500 **H** 13:487
G 500:487 **J** 487:13

8. What are the odds that you will end up on Go on your next roll in a game of Monopoly?

A 31:969 **C** 31:1000
B 969:31 **D** 1000:31

9. What are the odds that you will not end up on Park Place on your next roll in a game of Monopoly?

F 11:489 **H** 489:500
G 489:11 **J** 500:489

Name ______________________ Date __________ Class __________

LESSON 10-7

Reading Strategies

Use a Graphic Organizer

This graphic organizer will help you understand odds.

Definition of Odds A ratio between favorable outcomes and unfavorable outcomes	**Ways to Write Odds** Favorable to unfavorable Favorable:unfavorable Unfavorable to favorable Unfavorable:favorable
Odds in Favor A ratio of favorable outcomes to unfavorable outcomes. Of the 35 people who entered the contest, 12 could win a free CD. The odds in favor of winning a CD are 12 to 23.	**Odds Against** A ratio of unfavorable outcomes to favorable outcomes. Of the 35 people who entered the contest, 23 could not win a free CD. The odds against winning a CD are 23 to 12.

Odds

Use the information in the graphic organizer to help you answer the questions.

1. What is the meaning of "odds"?

____

2. What ratio is used to describe odds in favor?

____

3. What ratio is used to describe odds against?

____

4. What are the odds in favor of winning a CD?

____

5. What are the odds against winning a CD?

____

Name ______________________ Date __________ Class __________

Puzzles, Twisters & Teasers

Even Odder!

Solve the crossword puzzle. Use the formula *a*:*b* = odds in favor.

Across

2. The odds in ___ of an event is the ratio of favorable to unfavorable outcomes.

4. Odds and ___ are not the same thing.

5. a = number of ___ outcomes

9. If the probability of an event is $\frac{1}{3}$, this means that on average it will happen in one out of every three ___.

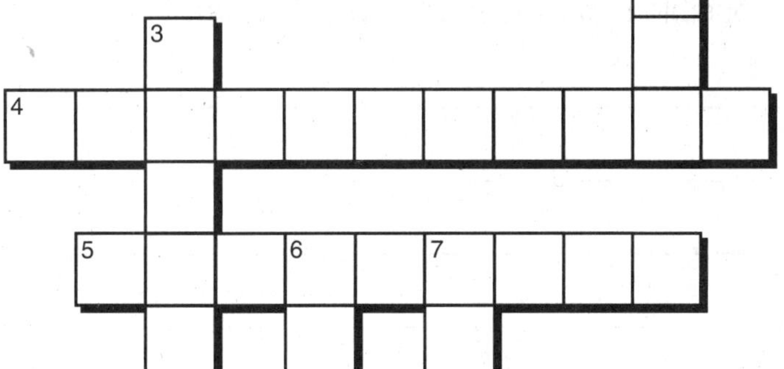

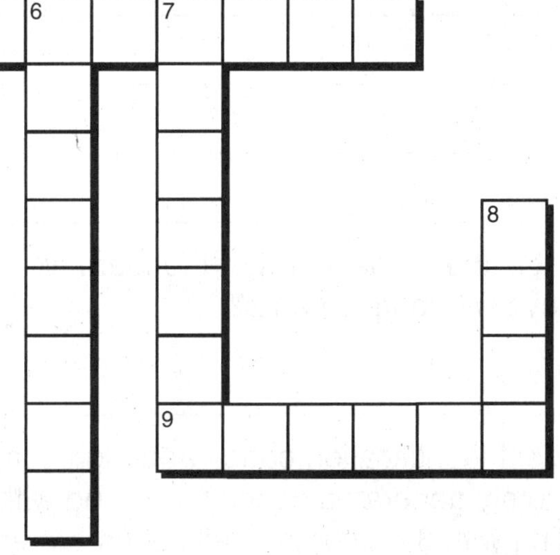

Down

1. It is possible to ___ odds to probabilities.

3. $a + b$ = ___ number of outcomes

6. b = number of unfavorable ___

7. The odds ___ an event is the ratio of unfavorable outcomes to favorable outcomes.

8. If the ___ in favor of an event are a:b, then the probability of the event occurring is $\frac{a}{a + b}$.

Name ______________________ Date ____________ Class ____________

LESSON 10-8

Practice A

Counting Principles

1. A snack bar serves tea, juice, and milk in small, medium, and large sizes. List all the different possible beverage orders.

____

____

____

2. The school's football team has a choice of different colored jerseys and different colored pants to wear for their uniforms. They have purple jerseys, white jerseys, and striped jerseys. The choices for the pants are purple or white. List all the different possible uniforms the team can wear.

____

____

____

____

____

____

3. What is the probability that the team will select the white jerseys with purple pants?

____

4. Student identification codes at a high school are 4-digit randomly generated codes beginning with 1 letter and ending with 3 numbers. If all codes are equally likely, how many possible codes are there?

____

5. Find the probability of being assigned the code A123.

____

6. Fabiana bought 3 fashion magazines, 2 exercise magazines, and 2 dance magazines. How many choices of magazines does she have to read?

____

Name ______________________ Date __________ Class __________

LESSON **10-8**

Practice B

Counting Principles

Employee identification codes at a company contain 2 letters followed by 2 numbers. All codes are equally likely.

1. Find the number of possible identification codes.

____

2. Find the probability of being assigned the code MT49.

____

3. Find the probability that an ID code of the company does not contain the letter *A* as the second letter of the code.

____

4. Find the probability that an ID code of the company does not contain the number 2.

____

5. Mrs. Sharpe is planning her dinners for next week. The choices for the entree are roast beef, turkey, or pork. The choices of carbohydrates are mashed potatoes, baked potatoes, or noodles. The vegetable choices are broccoli, spinach, or carrots. Make a tree diagram indicating the possible outcomes for each entree.

6. How many different meals could Mrs. Sharpe prepare? _______________

Find the probability for each of the following.

7. *P*(dinner with baked potato)

____

8. *P*(dinner with noodles and carrots)

____

9. Mitch bought 2 sports magazines, 3 guitar magazines, and 3 news magazines. How many choices of magazines does he have to read?

____

Name ______________________ Date ____________ Class ____________

LESSON **10-8**

Practice C

Counting Principles

Find the number of possible outcomes.

1. pasta: spaghetti, linguine

sauce: pesto, Alfredo, marinara

____

2. music: country, pop, rap

artist: male, female, duo, group

____

3. eye color: blue, brown, green

hair color: black, blond, brown, red

sex: male, female

____

4. font: Arial, Calligraphy, Helvetica

size: 10, 12, 14, 16, 20, 22, 24

color: black, red, blue, green

____

5. sport: baseball, basketball, football, hockey, soccer, volleyball

level: professional, college, high school, grade school _______________

Use the chart for Exercises 6 and 7.

Main Color	Trim Color	Frame Size	Tire Size	Gears
blue	white	19 in.	24 in.	15 speed
green	black	21 in.	26 in.	24 speed
red	yellow	23 in.		

6. Janis plans to buy a bike. How many combinations are possible with a choice of one main color, one trim color, one frame size, one tire size, and one gear selection? _______________

7. Janis decides to buy a green bike. How many combinations are now possible? _______________

A computer randomly generates a 5-character password of 3 letters followed by 2 digits. All passwords are equally likely.

8. Find the probability that a password contains exactly one 2. _______________

9. Find the probability that a password contains exactly one *A*. _______________

Name ______________________________ Date ____________ Class ____________

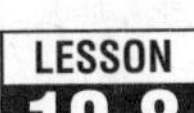

Reteach

Counting Principles

The Fundamental Counting Principle can help you solve some problems about situations that involve more than one activity.

the number of ways in which one activity can be performed × **the number of ways in which a second activity can be performed** = **the total number of ways in which both activities can be performed**

Apply the Fundamental Counting Principle to find the total number of possibilities in each situation.

1. Kelly has 6 shirts and 4 coordinating pants. The number of possible shirt-pants outfits is: ______________, or ________

2. The menu for dinner lists 2 soups, 4 meats, and 3 desserts. How many different meals that have one soup, one meat, and one dessert are possible? ______________, or ________

A **tree diagram** helps you see all the possibilities in a sample space.

If three coins are tossed at the same time, list all the possible outcomes.

List, in a column, the 2 possibilities for the 1st coin.

For each possibility for the 1st coin, list the 2 possibilities for the 2nd coin.

For each possibility for the 2nd coin, list the 2 possibilities for the 3rd coin.

Read the diagram across to write the list of all possible outcomes.

In this situation, there are $2 \times 2 \times 2 = 8$ possible outcomes.

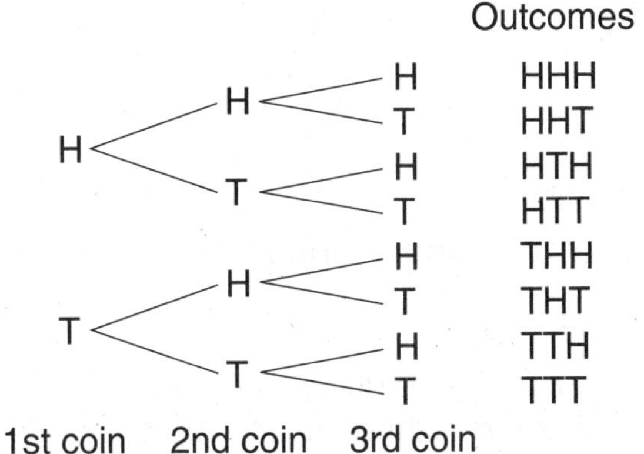

Draw a tree diagram and list the outcomes.

3. A vendor is selling cups of ice cream. There are 2 different sizes of cups: small (S), or large (L). There are 2 different flavors of ice cream: vanilla (V) or chocolate (C). There are 2 different toppings: fudge (F) or pineapple (P).

Name ______________________ Date ____________ Class ____________

LESSON 10-8

Reteach

Counting Principles (continued)

How many different 5-letter "words" are possible using the letters of TRIANGLE? Letters can be used only once in each "word."

There are 8 choices for the 1st letter, 7 choices for the 2nd letter, 6 choices for the 3rd letter, 5 choices for the 4th letter, and 4 choices for the 5th.

Apply the Fundamental Counting Principle.

$$\underset{\text{1st letter}}{8} \times \underset{\text{2nd letter}}{7} \times \underset{\text{3rd letter}}{6} \times \underset{\text{4th letter}}{5} \times \underset{\text{5th letter}}{4} = 6720 \text{ possibilities}$$

If a "word" is selected at random from the 6720 possibilities, what is the probability that it will be the "word" ANGLE?

There is only one outcome ANGLE. $P(\text{ANGLE}) = \frac{1}{6720}$

If a "word" is selected at random from the 6720 possibilities, what is the probability that it will not contain the letter G?

Find the number of favorable outcomes.

Eliminate the letter G from the choices. So, there are 7 choices to begin.

$$\underset{\text{1st letter}}{7} \times \underset{\text{2nd letter}}{6} \times \underset{\text{3rd letter}}{5} \times \underset{\text{4th letter}}{4} \times \underset{\text{5th letter}}{3} = 2520 \text{ possibilities}$$

$$P(\text{5-letter "word" with no G}) = \frac{\text{number of favorable outcomes}}{\text{total number of possible outcomes}} = \frac{2520}{6720}, \text{ or } \frac{3}{8}$$

Apply the Fundamental Counting Principle.

4. Consider the letters of the word MEDIAN.

a. How many different 4-letter "words" are possible? Letters can be used only once.

$\underset{\text{1st letter}}{\_\_\_\_} \times \underset{\text{2nd letter}}{\_\_\_\_} \times \underset{\text{3rd letter}}{\_\_\_\_} \times \underset{\text{4th letter}}{\_\_\_\_} =$ ______ possibilities

b. If a 4-letter "word" is selected at random from all the possibilities, then: $P(\text{DEAN}) =$ ______

c. If a 4-letter "word" is selected at random from all the possibilities, what is the probability that it will not contain the letter D?

favorable outcomes: $\underset{\text{1st letter}}{\_\_\_\_} \times \underset{\text{2nd letter}}{\_\_\_\_} \times \underset{\text{3rd letter}}{\_\_\_\_} \times \underset{\text{4th letter}}{\_\_\_\_} =$ ______

$P(\text{4-letter "word" with no D}) = \frac{\text{number of favorable outcomes}}{\text{total number of possible outcomes}} =$ ______, or ______

Name ______________________ Date __________ Class __________

LESSON **10-8**

Challenge

Answer the Phone

The world is divided into 9 telephone numbering zones. The North American Numbering Plan (NANP) was developed in 1947 to enable direct dialing without the need for an operator.

NANP numbers are 10 digits in length, of the form

N X X - N X X - X X X X
area code prefix line number

Originally, the plan created 86 areas and allowed for expansion to 144 areas. In 1995, NANP expanded to 792 area codes.

1. For the 3-digit area code NXX, the plan allows N to be any digit 2–9. Currently, there are no restrictions on the other 2 digits of the area code. How many area codes are possible?

   ____

2. For the 3-digit prefix, the plan allows N to be any digit 2–9. How many line numbers are possible for a given prefix?

   ____

3. How many telephone numbers are possible for a given area code?

   ____

Some of the prefixes are reserved for services.They are of the form N11 where N is any digit 2–9.

The most familiar service code is 911, reserved for emergency calls. Other currently assigned service codes are 411 (local directory assistance). 611 (repairs), 711 (teletypewriter [hearing/speech impaired]), 811(business office).

4. If all the service code prefixes are removed, how many telephone numbers are possible for a given area code?

   ____

Some other prefixes are not available for general use, such as:

555 (information), 800 and 888 (usually, but not always, toll free), 900 (pay per call).

5. For each prefix that is not available for general use, how many fewer telephone numbers are available for general use?

   ____

Name ________________________________ Date ___________ Class ___________

LESSON **10-8**

Problem Solving

Counting Principles

Write the correct answer.

1. The 5-digit zip code system for United States mail was implemented in 1963. How many different possibilities of zip codes are there with a 5-digit zip code where each digit can be 0 through 9?

____

2. In 1983, the ZIP +4 zip code system was introduced so mail could be more easily sorted by the 5-digit zip code plus an additional 4 digits. How many different possibilities of zip codes are there with the ZIP +4 system?

____

3. In Canada, each postal code has 6 symbols. The first, third and fifth symbols are letters of the alphabet and the second, fourth and sixth symbols are digits from 0 through 9. How many possible postal codes are there in Canada?

____

4. In the United Kingdom the postal code has 6 symbols. The first, second, fifth and sixth are letters of the alphabet and the third and fourth are digits from 0 through 9. How many possible postal codes are there in the United Kingdom?

____

Choose the letter for the best answer.

5. In Sharon Springs, Kansas, all of the phone numbers begin 852–4. The only differences in the phone numbers are the last 3 digits. How many possible phone numbers can be assigned using this system?

A 729 **C** 6561

B 1000 **D** 10,000

6. Many large cities have run out of phone numbers and so a new area code must be introduced. How many different phone numbers are there in a single area code if the first digit can't be zero?

F 90,000 **H** 9,000,000

G 4,782,969 **J** 10,000,000

7. How many different phone numbers are possible using a 3-digit area code and a 7-digit phone number if the first digit of the area code and phone number cannot be zero?

A 3,486,784,401 **C** 9,500,000,000

B 8,100,000,000 **D** 10,000,000,000

8. A shipping service offers to send packages by ground delivery using 2 different companies, by next day air using 3 different companies, and by 2-day air using 3 different companies. How many different shipping options does the service offer?

F 3 **H** 10

G 8 **J** 18

Name ______________________ Date __________ Class __________

Reading Strategies
Use a Visual Aid

If you have 2 pairs of shorts and 3 shirts, how many different outfits can you make?

Shorts A

Shorts B

Shirt A

Shirt B

Shirt C

A **tree diagram** is a way to visualize the possible outfits.

2 pairs of shorts → Each paired with one of the 3 shirts → Number of different outfits

Shorts A
- Shirt A → Shorts A & Shirt A
- Shirt B → Shorts A & Shirt B
- Shirt C → Shorts A & Shirt C

Shorts B
- Shirt A → Shorts B & Shirt A
- Shirt B → Shorts B & Shirt B
- Shirt C → Shorts B & Shirt C

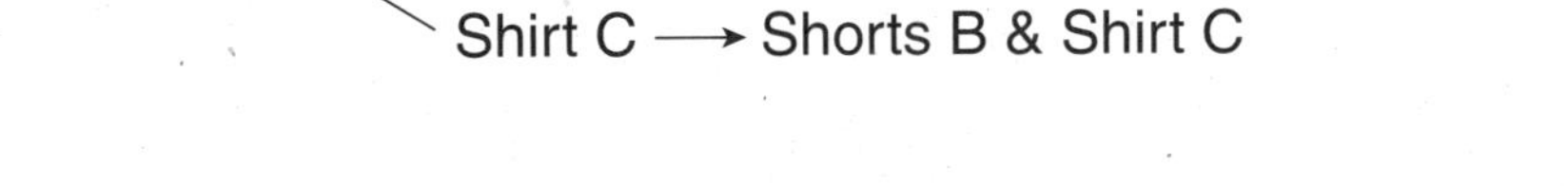

Use the tree diagram to answer the following questions.

1. How many different shirts could you pair with Shorts A?

____

2. How many different outfits could you make with Shorts A?

____

3. How many different shirts could you pair with Shorts B?

____

4. How many different outfits can you make with Shorts B?

____

5. With 2 different pairs of shorts and 3 different shirts, how many different outfits can you make?

____

Name ______________________________ Date ____________ Class ____________

LESSON 10-8

Puzzles, Twisters & Teasers

It's FUN-damental!

Determine if the number of possible outcomes is correct for each situation below. Circle the letter next to your answer. Use the letters to solve the riddle.

1. 3 types of birds and
2 types of cages
possible outcomes: 6

I correct	**A** incorrect

2. 4 colors and 3 sizes
possible outcomes: 7

M correct	**T** incorrect

3. 3 types of bagels and
3 types of spreads
possible outcomes: 9

J correct	**K** incorrect

4. 3 destinations and
4 months
possible outcomes: 34

P correct	**U** incorrect

5. 3 types of soup and
5 types of sandwiches
possible outcomes: 15

S correct	**U** incorrect

6. 5 destinations and
3 modes of transportation
possible outcomes: 8

X correct	**W** incorrect

7. 5 shirts, 3 pants, and 2 jackets
possible outcomes: 30

A correct	**Z** incorrect

8. 4 main dishes, 5 appetizers,
and 5 desserts
possible outcomes: 14

L correct	**V** incorrect

9. 3 types of paint and
1 type of paper
possible outcomes: 3

E correct	**B** incorrect

10. 10 type fonts, 5 type sizes,
and 3 paper sizes
possible outcomes: 18

V correct	**D** incorrect

What did the wave say to the beach?

Nothing, ___ ___ ___ ___ ___ T ___ ___ ___ ___ ___.

Name ____________________ Date __________ Class __________

LESSON **10-9**

Practice A

Permutations and Combinations

Express each expression as a product of factors.

1. $6!$ ____________________

2. $3!$ ____________________

3. $7!$ ____________________

4. $\frac{8!}{5!}$ ____________________

5. $\frac{4!}{2!}$ ____________________

6. $\frac{9!}{6!}$ ____________________

Evaluate each expression.

7. $5!$ __________

8. $9!$ __________

9. $3!$ __________

10. $8!$ __________

11. $\frac{7!}{4!}$ __________

12. $\frac{8!}{7!}$ __________

13. $\frac{5!}{2!}$ __________

14. $7! - 5!$ __________

15. $(6 - 3)!$ __________

16. $\frac{4!}{(6 - 2)!}$ __________

17. $\frac{9!}{(8 - 3)!}$ __________

18. $\frac{7!}{(9 - 4)!}$ __________

19. An anagram is a rearrangement of the letters of a word or words to make other words. How many possible arrangements of the letters W, O, R, D, and S can be made? ____________

20. Janell is having a group of friends over for dinner and is setting the name cards on the table. She has invited 5 of her friends for dinner. How many different seating arrangements are possible for Janell and her friends at the table? ____________

21. How many different selections of 4 books can be made from a bookcase displaying 12 books? ____________

Name ______________________ Date __________ Class __________

LESSON 10-9

Practice B

Permutations and Combinations

Evaluate each expression.

1. 10!

2. 13!

3. 11! − 8!

4. 12! − 9!

5. $\frac{15!}{8!}$

6. $\frac{18!}{12!}$

7. $\frac{13!}{(17 - 12)!}$

8. $\frac{19!}{(15 - 2)!}$

9. $\frac{15!}{(18 - 10)!}$

10. Signaling is a means of communication through signals or objects. During the time of the American Revolution, the colonists used combinations of a barrel, basket, and a flag placed in different positions atop a post. How many different signals could be sent by using 3 flags, one above the other on a pole, if 8 different flags were available?

11. From a class of 25 students, how many different ways can 4 students be selected to serve in a mock trial as the judge, defending attorney, prosecuting attorney, and the defendant?

12. How many different 4 people committees can be formed from a group of 15 people?

13. The girls' basketball team has 12 players. If the coach chooses 5 girls to play at a time, how many different teams can be formed?

14. A photographer has 50 pictures to be placed in an album. How many combinations will the photographer have to choose from if there will be 6 pictures placed on the first page?

Name ______________________ Date __________ Class __________

LESSON 10-9

Practice C

Permutations and Combinations

Evaluate each expression.

1. $\frac{16!}{(15 - 4)!}$ ____________

2. $\frac{21!}{(19 - 3)!}$ ____________

3. $\frac{17!}{5!(17 - 5)!}$ ____________

4. ${}_7P_3$ ____________

5. ${}_9P_4$ ____________

6. ${}_{10}P_8$ ____________

7. ${}_{18}P_2$ ____________

8. ${}_9C_2$ ____________

9. ${}_{11}C_5$ ____________

10. ${}_{13}C_{11}$ ____________

11. ${}_{15}C_3$ ____________

12. The music class has 20 students and the teacher wants them to practice in groups of 5. How many different ways can the first group of 5 be chosen?

____

13. Math, science, English, history, health, and physical education are the subjects on Jamar's schedule for next year. Each subject is taught in each of the 6 periods of the day. From how many different schedules will Jamar be able to choose?

____

14. The Hamburger Trolley has 25 different toppings available for their hamburgers. They have a $3 special that is a hamburger with your choice of 5 different toppings. Assume no toppings are used more than once. How many different choices are available for the special?

____

15. Many over the counter stocks are traded through Nasdaq, an acronym for the National Association of Securities Dealers Automatic Quotations. Most of the stocks listed on the Nasdaq use a 4-digit alphabetical code. For example, the code for Microsoft is MSFT. How many different 4-digit alphabetical codes could be available for use by the association? Assume letters cannot be reused.

____

Name ______________________ Date __________ Class __________

LESSON 10-9

Reteach

Permutations and Combinations

Factorial: a string of factors that counts down to 1

$6! = 6 \cdot 5 \cdot 4 \cdot 3 \cdot 2 \cdot 1$

To evaluate an expression with factorials, cancel common factors.

$\frac{5!}{3!} = \frac{5 \cdot 4 \cdot \cancel{3} \cdot \cancel{2} \cdot \cancel{1}}{\cancel{3} \cdot \cancel{2} \cdot \cancel{1}} = 5 \cdot 4 = 20$

Complete to evaluate each expression.

1. $\frac{7!}{4!} = \frac{7 \cdot 6 \cdot 5 \cdot \_ \cdot \_ \cdot \_ \cdot \_}{\_\_\_\_\_\_}$

$= 7 \cdot$ ______ = ______

2. $\frac{6!}{(5-2)!} = \frac{6!}{\_!} = \frac{\_ \cdot \_ \cdot \_ \cdot \_ \cdot \_ \cdot \_}{\_ \cdot \_ \cdot \_}$

= ______ = ______

Permutation: an arrangement in which order is important

wxyz is not the same as *yxzw*

Apply the Fundamental Counting Principle to find how many permutations are possible using all 4 letters *w*, *x*, *y*, *z* with no repetition.

$\frac{4}{\text{1st letter}} \times \frac{3}{\text{2nd letter}} \times \frac{2}{\text{3rd letter}} \times \frac{1}{\text{4th letter}} = 4! = 24$ possible arrangements

When you arrange *n* things, *n*! permutations are possible.

Complete to find the number of permutations.

3. In how many ways can 6 people be seated on a bench that seats 6?

$6! =$ ___ • ___ • ___ • ___ • ___ • ___

= ______ possibilities

4. How many 5-digit numbers can be made using the digits 7, 4, 2, 1, 8 without repetitions?

___! = ___ • ___ • ___ • ___ • ___

= ______ possibilities

Apply the Fundamental Counting Principle to find how many permutations are possible using 4 letters 2 at a time, with no repetitions.

$\frac{4}{\text{1st letter}} \times \frac{3}{\text{2nd letter}} = 12$ possible 2-letter arrangements

Apply the Foundamental Counting Principle.

5. In how many ways can 6 people be seated on a bench that seats 4?

$\frac{\_\_\_}{\text{1st seat}} \times \frac{\_\_\_}{\text{2nd seat}} \times \frac{\_\_\_}{\text{3rd seat}} \times \frac{\_\_\_}{\text{4th seat}} =$ ______ possibilities

6. How many 3-digit numbers can be made using the digits 7, 4, 2, 1, 8 without repetitions?

$\frac{\_\_\_}{\text{1st digit}} \times \frac{\_\_\_}{\text{2nd digit}} \times \frac{\_\_\_}{\text{3rd digit}} =$ ______ possibilities

Name ______________________ Date __________ Class __________

LESSON 10-9

Reteach

Permutations and Combinations (continued)

When using fewer than the available number of items in an arrangement, instead of the Fundamental Counting Principle, you can use a formula to find the number of possible permutations.

To arrange *n* things *r* at a time, the number of possible permutations *P* is: $_nP_r = \frac{n!}{(n-r)!}$

Find how many permutations are possible using 4 letters 2 at a time, with no repetitions.

$$_4P_2 = \frac{4!}{(4-2)!} = \frac{4!}{2!} = \frac{4 \cdot 3 \cdot \cancel{2} \cdot \cancel{1}}{\cancel{2} \cdot \cancel{1}} = 12 \text{ possible 2-letter arrangements}$$

Complete to apply the permutations formula.

7. In how many ways can 6 people be seated on a bench that seats 4?

$$_6P_4 = \frac{6!}{(6 - \quad)!} = \frac{\quad !}{\quad !} = \underline{\hspace{6cm}} = \underline{\hspace{1.5cm}} \text{ possible seating arrangements}$$

8. How many 3-digit numbers can be made using the digits 7, 4, 2, 1, 8 without repetitions?

$$_5P_{\underline{\hspace{1.5cm}}} = \frac{5!}{(5 - \quad)!} = \frac{\quad !}{\quad !} = \underline{\hspace{6cm}} = \underline{\hspace{1.5cm}} \text{ possible 3-digit numbers}$$

Combination: an arrangement in which order is not important

How many 2-letter combinations can be made from the 4 letters *w*, *x*, *y*, *z* without repetition?

wx	~~*xw*~~	~~*yw*~~	~~*zw*~~
wy	*xy*	~~*yx*~~	~~*zx*~~
wz	*xz*	*yz*	~~*zy*~~

There are fewer combinations than permutations.

The combinations *w x* and *x w* are the same. After all the same combinations are removed, there are 6 different combinations possible.

$$_nC_r = \frac{_nP_r}{r!} = \frac{n!}{(n-r)!\, r!}$$

The number of combinations *C* of *n* things taken *r* at a time is:

$$_4C_2 = \frac{_4P_2}{2!} = \frac{4!}{(4-2)!\,2!} = \frac{4!}{2!\,2!} = \frac{\cancel{4}^2 \cdot 3 \cdot \cancel{2} \cdot \cancel{1}}{\cancel{2} \cdot 1 \cdot \cancel{2} \cdot \cancel{1}} = 2 \cdot 3 = 6$$

Complete to apply the combinations formula.

9. How many different 4-person committees can be formed from a group of 6 people?

$$_6C_4 = \frac{_6P_4}{4!} = \frac{6!}{(6 - \quad)!\quad !} = \frac{6!}{\quad !\quad !} = \underline{\hspace{6cm}} = \underline{\hspace{1.5cm}} \text{ possible 4-person committees}$$

Name ______________________ Date __________ Class __________

LESSON 10-9

Challenge

Roundtable Discussion

The number of ways in which 4 people can be seated *in a row*, on a bench that seats 4 is ${}_4P_4$, or 4!.

$$\frac{4}{\text{1st seat}} \times \frac{3}{\text{2nd seat}} \times \frac{2}{\text{3rd seat}} \times \frac{1}{\text{4th seat}} = 4! = {}_4P_4 = 24 \text{ different arrangements}$$

Now consider what happens if 4 people are seated *in a circle*,

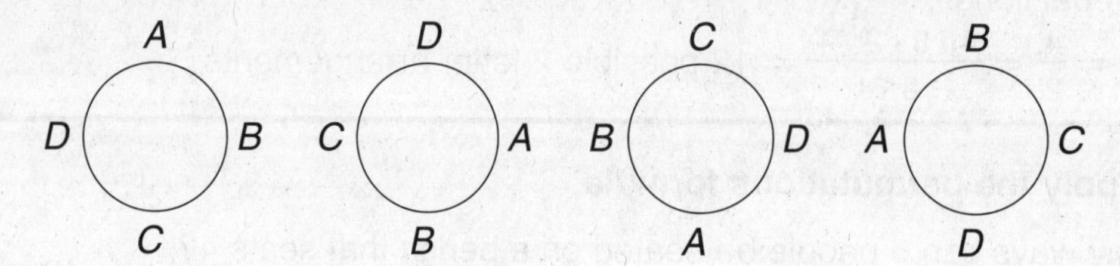

around a round table that seats 4.

Note that the 4 circular arrangements shown are really all the same with respect to who sits next to whom.

For each of the 4! permutations, there are 4 alike.

So, there are fewer ways to seat 4 people at a circular table that seats 4.

$$\frac{{}_4P_4}{4} = \frac{4!}{4} = \frac{4 \cdot 3 \cdot 2 \cdot 1}{4} = 6 \text{ different arrangements}$$

1. In how many different ways can 5 people be seated in a row, on a bench that seats 5?

____

2. In how many different ways can 5 people be seated in a circle, around a circular table that seats 5?

____

3. In how many different ways can *n* people be seated in a row, on a bench that seats *n*? Answer in factorial form.

____

4. In how many different ways can *n* people be seated in a circle, around a circular table that seats *n*? Answer in factorial form.

____

Name ______________________ Date __________ Class __________

Problem Solving

Permutations and Combinations

Write the correct answer.

1. In a day camp, 6 children are picked to be team captains from the group of children numbered 1 through 49. How many possibilities are there for who could be the 6 captains?

   ____

2. If you had to match 6 players in the correct order for most popular outfielder from a pool of professional players numbered 1 through 49, how many possibilities are there?

   ____

Volleyball tournaments often use pool play to determine which teams will play in the semi-final and championship games. The teams are divided into different pools, and each team must play every other team in the pool. The teams with the best record in pool play advance to the final games.

3. If 12 teams are divided into 2 pools, how many games will be played in each pool?

   ____

4. If 12 teams are divided into 3 pools, how many pool play games will be played in each pool?

   ____

A word jumble game gives you a certain number of letters that you must make into a word. Choose the letter for the best answer.

5. How many possibilities are there for a jumble with 4 letters?

   A 4 **C** 24
   B 12 **D** 30

6. How many possibilities are there for a jumble with 5 letters?

   F 24 **H** 120
   G 75 **J** 150

7. How many possibilities are there for a jumble with 6 letters?

   A 120
   B 500
   C 720
   D 1000

8. On the Internet, a site offers a program that will un-jumble letters and give you all of the possible words that can be made with those letters. However, the program will not allow you to enter more than 7 letters due to the amount of time it would take to analyze. How many more possibilities are there with 8 letters than with 7?

   F 5040 **G** 20,640
   H 35,280 **J** 40,320

Name ______________________ Date __________ Class __________

LESSON 10-9

Reading Strategies

Use a Visual Aid

A **permutation** is an arrangement of objects in a certain order.

How many different ways can you arrange these three shapes?

You can use a tree diagram to visualize all of the possible arrangements:

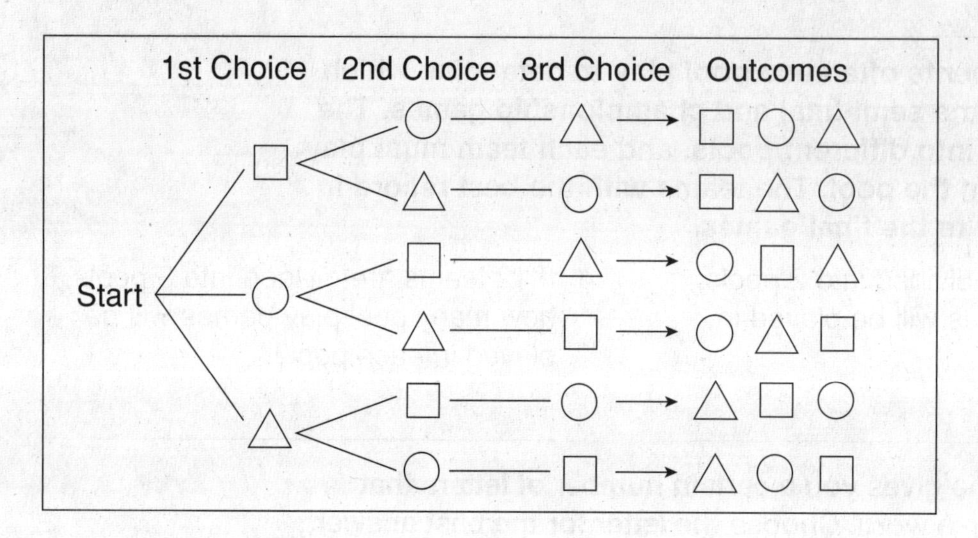

Use the tree diagram to answer the following.

1. If you start with the circle, how many different arrangements can you make? List them.

____

2. If you start with the square, how many different arrangements can you make? List them.

____

3. If you start with the triangle, how many different arrangements can you make? List them.

____

4. How many different arrangements can you make with these three shapes?

____

Name ______________________ Date ____________ Class ____________

LESSON **10-9**

Puzzles, Twisters & Teasers

Finding a Treasure!

Black out the incorrect expressions to see a shape.

9! = 362,880	2! = 4	11! = 9,497,876	3! = 9	5! = 120
5! = 25	3! = 6	4! = 14	2! = 2	10! = 100
6! = 150	5! = 125	8! = 40,320	7! = 49	6! = 36
4! = 256	7! = 5040	10! = 10,000,000	10! = 3,628,8000	7! = 823,543
4! = 24	9! = 729	12! = 144	8! = 16,777,216	11! = 39,916,800

What do you see? ______________________

LESSON 10-1

Practice A
Probability

1. Meteorology is the study of the atmosphere, natural phenomenon, atmospheric conditions, weather and climate. A meteorologist forecasts and reports the weather. A meteorologist forecasts a 70% chance of rain. What is the probability of each outcome?

Outcome	Rain	No rain
Probability	0.7 or 70%	0.3 or 30%

Use the spinner to determine the probability of each outcome.

2. $P(1)$ $\frac{1}{4}$
3. $P(4)$ $\frac{1}{4}$
4. $P(\text{even number})$ $\frac{1}{2}$
5. $P(5)$ 0
6. $P(\text{odd number})$ $\frac{1}{2}$
7. $P(\text{an integer})$ 1
8. $P(1 \text{ or } 2)$ $\frac{1}{2}$
9. $P(\text{number} > 1)$ $\frac{3}{4}$
10. $P(\text{a whole number})$ 1

11. Mrs. Silverstein has 14 boys in her class of 25 students. She must select one student at random to serve as the class moderator. What is the probability that she will choose a boy? What is the probability that she will choose a girl?

$P(\text{boy}) = \frac{14}{25}$; $P(\text{girl}) = \frac{11}{25}$

12. When tossing a regular coin, what is the probability of it landing on heads?

$\frac{1}{2}$

LESSON 10-1

Practice B
Probability

These are the results of the last math test. The teacher determines that anyone with a grade of more than 70 passed the test. Give the probability for the indicated grade.

Grade	65	70	80	90	100
# of Students	5	3	12	10	2

1. $P(70)$ $\frac{3}{32}$
2. $P(100)$ $\frac{1}{16}$
3. $P(80)$ $\frac{3}{8}$
4. $P(\text{passing})$ $\frac{3}{4}$
5. $P(\text{grade} > 80)$ $\frac{3}{8}$
6. $P(60)$ 0
7. $P(\text{failing})$ $\frac{1}{4}$
8. $P(\text{grade} \le 80)$ $\frac{5}{8}$

A bowling game consists of rolling a ball and knocking up to 5 pins down. The number of pins knocked down are then counted. The table gives the probability of each outcome.

Number of Pins Down	0	1	2	3	4	5
Probability	0.175	0.189	0.264	0.205	0.132	0.035

9. What is the probability of knocking down all 5 pins?
0.035, or 3.5%

10. What is the probability of knocking down no pins?
0.175, or 17.5%

11. What is the probability of knocking down at most 2 pins?
0.628, or 62.8%

12. What is the probability of knocking down at least 2 pins?
0.636, or 63.6%

13. What is the probability of knocking down more than 3 pins?
0.167, or 16.7%

LESSON 10-1

Practice C
Probability

Demographers often use statistics to predict and explain future changes in populations in many areas including housing, education, life events, and unemployment. A demographer developed this chart to illustrate the cause of death in his community of 50,000 people.

Give the probability for each outcome.

Outcome	Heart	Cancer	Accident	Respiratory	Other
Probablility	0.35	0.28	0.16	0.13	0.08

1. $P(\text{death from cancer})$ $\frac{7}{25}$
2. $P(\text{death from accident})$ $\frac{4}{25}$
3. $P(\text{death from heart})$ $\frac{7}{20}$
4. $P(\text{non-accidental death})$ $\frac{21}{25}$
5. $P(\text{death from other})$ $\frac{2}{25}$
6. $P(\text{death from heart or cancer})$ $\frac{63}{100}$

Use the spinner to determine the probability of each outcome.

7. $P(\text{white 1})$ $\frac{1}{8}$
8. $P(\text{dots 2})$ $\frac{1}{4}$
9. $P(\text{lines even})$ $\frac{1}{8}$
10. $P(\text{dots 1})$ $\frac{1}{8}$
11. $P(\text{white odd})$ $\frac{3}{8}$
12. $P(\text{dots integer})$ $\frac{3}{8}$
13. $P(\text{odd})$ $\frac{5}{8}$
14. $P(\text{white or 2})$ $\frac{3}{4}$

15. There are six teams competing to collect the most food for the food bank. Team B has a 30% chance of winning. Teams A, C, D, and E all have the same chance of winning. Team F is one third as likely to win as Team B. Create a table of probabilities for the sample space.

Outcome	Team A	Team B	Team C	Team D	Team E	Team F
Probablility	0.15	0.3	0.15	0.15	0.15	0.1

LESSON 10-1

Reteach
Probability

The **probability** that something will happen is how often you can expect that **event** to occur. This depends upon how many outcomes are possible, the **sample space.**

In the spinner shown, the circle is divided into four equal parts. There are 4 possible outcomes.

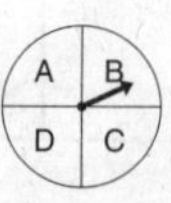

So, in a single spin:

$P(A) = P(B) = P(C) = P(D) = 25\% = \frac{1}{4}$

Complete to give the probability for each event.

	1. A fair coin is tossed.	2. A number cube is rolled.
List all the possible outcomes.	heads, tails	1, 2, 3, 4, 5, 6
How many outcomes in sample space?	2	6
Find the probability of the event shown.	$P(\text{heads}) = \frac{1}{2}$	$P(5) = \frac{1}{6}$

- A probability of 0 means the event is **impossible,** or can never happen.
  On the spinner above, $P(F) = 0$.
- A probability of 1 means the event is **certain,** or has to happen.
  In one roll of a number cube, $P(\text{a whole number from 1 through 6}) = 1$.

Give the probability for each event.

3. selecting a rectangle from the set of squares
$P(\text{rectangle}) =$ 1

4. selecting a negative number from the set of whole numbers
$P(\text{negative number}) =$ 0

- The sum of the probabilities of all the possible outcomes in a sample space is 1.
  If the probability of *snow* is 30%, then the probability of *no snow* is 70%.
  $P(\text{snow}) + P(\text{no snow}) = 1$

5. If the probability of selecting a senior for a committee is 60%, then the probability of not selecting a senior is:
40%

6. If the probability of choosing a red ball from a certain box is 0.35, then the probability of not choosing a red ball is:
0.65

LESSON 10-1
Reteach
Probability (continued)

To find the probability that an event will occur, add the probabilities of all the outcomes included in the event.

This bar graph shows the midterm grades of the 30 students in Ms. Lin's class.

What is the probability that Susan has a grade of C or higher?

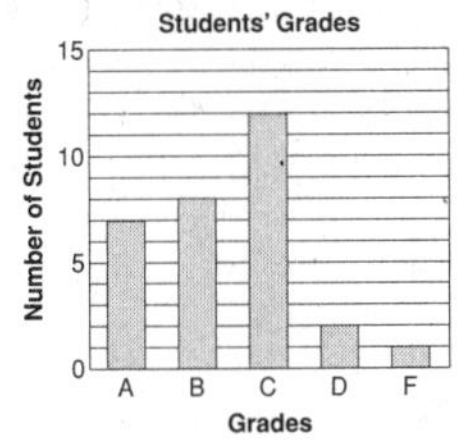

The event "a grade of C or higher" consists of the outcomes C, B, A.

$P(\text{C or higher}) = P(\text{C}) + P(\text{B}) + P(\text{A})$

$= \frac{12}{30} + \frac{8}{30} + \frac{7}{30} = \frac{12 + 8 + 7}{30} = \frac{27}{30} = \frac{9}{10} = 90\%$

So, the probability that Susan's midterm grade is C or higher is $\frac{9}{10}$ or 90%.

Use the bar graph above to find each probability.

7. B or higher is honor roll. What is the probability that Ken made the honor roll?

   $P(\text{B or higher}) = P(\text{B}) + \underline{P(\text{A})}$

   $= \frac{8}{30} + \frac{7}{30}$

   $= \frac{15}{30} = \underline{\frac{1}{2}}$

   So, the probability that Ken made the honor roll is: $\underline{\frac{1}{2}}$ or $\underline{50}$ %.

8. In this class, D is a failing grade. What is the probability that Tom failed?

   $P(\text{D or lower}) = P(\text{D}) + \underline{P(\text{F})}$

   $= \frac{2}{30} + \frac{1}{30}$

   $= \frac{3}{30} = \underline{\frac{1}{10}}$

   So, the probability that Tom failed is: $\underline{\frac{1}{10}}$ or $\underline{10}$ %.

LESSON 10-1
Challenge
Why We Look Like Our Parents

Each parent carries two genes with respect to a specific trait and each passes one of these genes on to an offspring who then also has two genes for that trait.

In pea plants, a tall gene is dominant over a short gene. So, if a pea plant has at least one tall gene, the plant is tall. If T represents *tall* and t represents *short*, one way to represent the gene makeup with respect to height of a tall pea plant would be Tt.

1. What is another way to represent the gene makeup with respect to height of a tall pea plant? TT

An early 20th-century English geneticist, Reginald Punnett, invented a method to display the gene makeup of parents and their offspring.

2. The **Punnett square** at the right shows the gene makeup of one tall parent plant as the labels for the columns. Insert your result from Question 1 for the other tall parent plant as the row labels.

	T	t
T		
T		

3. a. The label for each column has been inserted in each box of its column, as the first gene of the offspring plant. Insert your labels for each row as the second gene of each offspring plant.

   b. According to the two genes now in each of the boxes for the new offspring plants, tell if the new plant will be tall or short.

   c. What is the probability that an offspring of these tall parent plants will be tall?

   $\frac{4}{4}$ or 1 or 100%

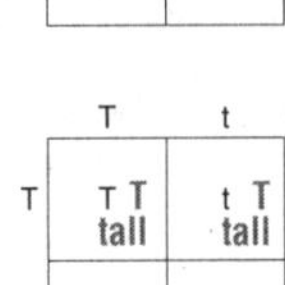

4. Suppose the gene makeup for both tall parent pea plants is Tt.

   a. Complete a Punnett square to display the gene makeup of the offspring.

   b. What is the probability that an offspring of these tall parent plants will be tall?

   $\frac{3}{4}$ or 75%

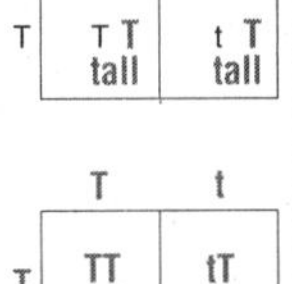

LESSON 10-1
Problem Solving
Probability

Write the correct answer.

1. To get people to buy more of their product, a company advertises that in selected boxes of their popsicles is a super hero trading card. There is a $\frac{1}{4}$ chance of getting a trading card in a box. What is the probability that there will not be a trading card in the box of popsicles that you buy?

   $\frac{3}{4}$

2. The probability of winning a lucky wheel television game show in which 6 preselected numbers are spun on a wheel numbered 1–49 is $\frac{1}{13{,}983{,}816}$ or 0.000007151%. What is the probability that you will not win the game show?

   $\frac{13{,}983{,}815}{13{,}983{,}816}$

Based on world statistics, the probability of identical twins is 0.004, while the probability of fraternal twins is 0.023.

3. What is the probability that a person chosen at random from the world will be a twin?

   0.027, or 2.7%

4. What is the probability that a person chosen at random from the world will not be a twin?

   0.973, or 97.3%

Use the table below that shows the probability of multiple births by country. Choose the letter for the best answer.

Probability of Multiple Births

Country	Probability
Canada	0.012
Japan	0.008
United Kingdom	0.014
United States	0.029
Sweden	0.014
Switzerland	0.013

5. In which country is it most likely to have multiple births?
   A Japan
   C Sweden
   (B) United States
   D Switzerland

6. In which country is it least likely to have multiple births?
   (F) Japan
   H Sweden
   G United States
   J Switzerland

7. In which two countries are multiple births equally likely?
   A United Kingdom, Canada
   B Canada, Switzerland
   (C) Sweden, United Kingdom
   D Japan, United States

LESSON 10-1
Reading Strategies
Focus On Vocabulary

Probability is the chance that something will happen.

An **event** has an outcome that can be stated using probability. The probability of something happening is written ***P*(event).**

The probability of an event happening is described by a number between 0 and 1, or as a percent between 0% and 100%.

A probability of 0 or 0% means it is **impossible** for the event to happen.

A probability of 1 or 100% means it is **certain** the event will happen.

Answer each question.

1. What name is given to the chance of something happening?

   probability

2. How is the probability of an event written?

   P(event)

3. What number or percent is used to mean that an event is certain?

   1 or 100%

An **experiment** is an activity where results are observed. Flipping a coin and tossing a number cube are both experiments.

In an experiment, each observation (such as one coin toss) is called a **trial.** The result of the trial is called an **outcome** (such as a coin landing on heads).

Use the terms above to answer the following questions.

4. What do you call an activity in which results are observed?

   an experiment

5. What is one observation in an experiment called?

   a trial

6. What is an outcome?

   the result of a trial

LESSON 10-1

Puzzles, Twisters & Teasers

What Are Your Chances?

Solve the crossword puzzle.

Across

3. A probability of 0 means the event is ___. IMPOSSIBLE
5. The probabilities of all the outcomes in the sample space add up to ___. ONE
7. An ___ is any set of one or more outcomes. EVENT
9. Each observation in an experiment is called a ___. TRIAL

Down

1. Each result of an experiment is called an ___. OUTCOME
2. A probability of 1 means an event is ___. CERTAIN
4. The ___ of an event is a number that tells how likely the event is to happen. PROBABILITY
6. An ___ is an activity in which results are observed. EXPERIMENT
8. The set of all possible outcomes of an experiment is the ___ space. SAMPLE

LESSON 10-2

Practice A

Experimental Probability

The results of an unbiased survey show the favorite instruments of 8th graders. Estimate the probability of each.

Result	Piano	Drums	Trombone	Flute	Violin	Clarinet
Number	1	4	42	38	12	3

1. a student chooses clarinet $\frac{3}{100}$ or 3%
2. a student chooses drums $\frac{1}{25}$ or 4%
3. a student chooses flute $\frac{19}{50}$ or 38%
4. a student chooses piano $\frac{1}{100}$ or 1%
5. a student chooses trombone $\frac{21}{50}$ or 42%
6. a student chooses violin $\frac{3}{25}$ or 12%

A can contains color chips in 5 different colors. Thomas took a sample from the can and counted the colors. His results are in the table below.

Color	Blue	Pink	Black	White	Green
Number	10	5	20	30	15

7. Use the table to compare the probability that Thomas chooses a pink color chip to the probability that he chooses a white color chip.
   0.0625; 0.375; less likely
8. Use the table to compare the probability that Thomas chooses a green color chip to the probability that he chooses a blue color chip.
   0.1875; 0.125; more likely
9. Cheryl surveyed 30 students who ride the bus to school, 8 who walk, 9 who ride bicycles, and 3 who ride in cars. Estimate the probability that the next student Cheryl surveys will walk to school. $\frac{4}{25}$ or 16%

LESSON 10-2

Practice B

Experimental Probability

1. A number cube was thrown 150 times. The results are shown in the table below. Estimate the probability for each outcome.

Outcome	1	2	3	4	5	6
Frequency	33	21	15	36	27	18
Probability	22%	14%	10%	24%	18%	12%

A movie theater sells popcorn in small, medium, large and jumbo sizes. The customers of the first show purchase 4 small, 20 medium, 40 large, and 16 jumbo containers of popcorn. Estimate the probability of the purchase of each of the different size containers of popcorn.

2. P(small container) $\frac{1}{20}$ or 5%
3. P(medium container) $\frac{1}{4}$ or 25%
4. P(large container) $\frac{1}{2}$ or 50%
5. P(jumbo container) $\frac{1}{5}$ or 20%

Janessa polled 154 students about their favorite winter sport.

Outcome	Frequency
Skiing	46
Sledding	21
Snowboarding	64
Ice Skating	14
Other	9

6. Use the table to compare the probability that a student chose snowboarding to the probability that a student chose skiing.
   ≈0.415; ≈0.299; more likely
7. Use the table to compare the probability that a student chose ice skating to the probability that a student chose sledding.
   ≈0.091; ≈0.136; less likely
8. The class president made 75 copies of the flyer advertising the school play. It was found that 8 of the copies were defective. Estimate the probability that a flyer will be printed properly. ≈0.893

LESSON 10-2

Practice C

Experimental Probability

The developer of a Web page wants to track the number of hits to each link of the Web page. An automatic counter records the following hits in one week: home, 60 hits; FAQ, 20 hits; employment opportunities, 15 hits; products, 50 hits; order status, 30 hits; and contact information, 25 hits. Estimate the probability of each.

1. P(home) $\frac{3}{10}$ or 30%
2. P(FAQ) $\frac{1}{10}$ or 10%
3. P(products) $\frac{1}{4}$ or 25%
4. P(order status) $\frac{3}{20}$ or 15%
5. P(employment opportunities) $\frac{3}{40}$ or 7.5%
6. P(contact information) $\frac{1}{8}$ or 12.5%

Hayley bought a CD with 12 songs on it. She placed it in her CD changer and selected random play mode. Hayley kept a record of how the tracks played. The following table illustrates the results.

Track	1	2	3	4	5	6	7	8	9	10	11	12
Frequency	2	4	3	1	2	4	2	3	5	2	1	4

Estimate the probability for each of the following.

7. P(track 2) ≈0.1212
8. P(track 4) ≈0.0303
9. P(track 5) ≈0.0606
10. P(track 8) ≈0.0909
11. P(track 9) ≈0.1515
12. P(track 13) 0
13. Use the table to compare the probability that Track 10 was played to the probability that Track 6 was played.
    ≈0.0606; ≈0.1212; ≈less likely
14. A coin is tossed 70 times, and it lands on heads 36 times. Estimate the probability of it landing on tails. ≈0.4857

LESSON 10-2

Reteach

Experimental Probability

A machine is filling boxes of apples by choosing 50 apples at random from a selection of six types of apples. An inspector records the results for one filled box in the table below.

Type	Pink Lady	Red Delicious	Granny Smith	Golden Delicious	Fuji	MacIntosh
Number	8	12	6	4	15	5

The inspector then expands the table to find the experimental probability.

$$\text{probability} = \frac{\text{number of type of apple}}{\text{total number of apples}}$$

Type	Pink Lady	Red Delicious	Granny Smith	Golden Delicious	Fuji	MacIntosh
Experimental Probability (ratio)	$\frac{8}{50}$, or $\frac{4}{25}$	$\frac{12}{50}$, or $\frac{6}{25}$	$\frac{6}{50}$, or $\frac{3}{25}$	$\frac{4}{50}$, or $\frac{2}{25}$	$\frac{15}{50}$, or $\frac{3}{10}$	$\frac{5}{50}$, or $\frac{1}{10}$
Experimental Probability (percent)	16%	24%	12%	8%	30%	10%

Find each sum for the apple experiment.

1. The sum of the experimental probability ratios.

   $\text{probability} = \frac{8}{50} + \frac{12}{50} + \frac{6}{50} + \frac{4}{50} + \frac{15}{50} + \frac{5}{50} = \frac{50}{50}$ or 1

2. The sum of the experimental probability percents.

   probability = 16% + 24% + 12% + 8% + 30% + 10% = 100% or 1

Complete the table to find the experimental probability.

3. Five types of seed are inserted at random in a pre-seeded strip ready for planting.

Type	Marigold	Impatiens	Snapdragon	Daisy	Petunia
Number	40	100	80	60	120
Experimental Probability (ratio)	$\frac{40}{400}$, or $\frac{1}{10}$	$\frac{100}{400}$, or $\frac{1}{4}$	$\frac{80}{400}$, or $\frac{1}{5}$	$\frac{60}{400}$, or $\frac{3}{20}$	$\frac{120}{400}$, or $\frac{3}{10}$
Experimental Probability (percent)	10%	25%	20%	15%	30%

LESSON 10-2

Challenge

Tossing and Spinning

The more times you repeat an experiment, the closer the experimental probabiltiy and the theoretical probability become.

Toss a penny 200 times.

Heads	Tails

1. Record your results in the table. Results will vary.
2. What is the theoretical probability of:
   getting heads? $\frac{1}{2}$ getting tails? $\frac{1}{2}$
3. What is your experimental probability of:
   getting heads? near $\frac{1}{2}$ getting tails? near $\frac{1}{2}$
4. How close are your experimental probabilities to the theoretical probabilities?
   Possible answer: very close

Spin a penny 200 times.

Heads	Tails

5. Record your results in the table. Results will vary.
6. What is your experimental probability of:
   getting heads? $\frac{1}{2}$ getting tails? $\frac{1}{2}$
7. Compare your experimental probabilities for tossing the penny and spinning the penny. Are they close? Explain.
   Answers will vary.

Roll a number cube 200 times.

1	2	3	4	5	6

8. Record your results in the table. Results will vary.
9. What is the theoretical probability of:
   getting a 1? $\frac{1}{6}$ a 2? $\frac{1}{6}$ a 3? $\frac{1}{6}$ a 4? $\frac{1}{6}$ a 5? $\frac{1}{6}$ a 6? $\frac{1}{6}$
10. What is your experimental probability of getting:
    a 1? vary a 2? vary a 3? vary a 4? vary a 5? vary a 6? vary
11. How close are your experimental probabilities to the theoretical probabilities?
    Possible answer: very close

LESSON 10-2

Problem Solving

Experimental Probability

Use the table below. Round to the nearest percent. Write the correct answer.

Average Number of Days of Sunshine Per Year for Selected Cities

City	Number of Days
Buffalo, NY	175
Fort Wayne, IN	215
Miami, FL	256
Raleigh, NC	212
Richmond, VA	230

1. Estimate the probability of sunshine in Buffalo, NY.
   48%
2. Estimate the probability of sunshine in Fort Wayne, IN.
   59%
3. Estimate the probability of sunshine in Miami, FL.
   70%
4. Estimate the probability that it will not be sunny in Raleigh, NC.
   42%
5. Estimate the probability that it will not be sunny in Miami, FL.
   30%
6. Estimate the probability of sunshine in Richmond, VA.
   63%

Use the table below that shows the number of deaths and injuries caused by lightning strikes. Choose the letter for the best answer.

States with Most Lightning Deaths

State	Average deaths per year	Average injuries per year	Population
Florida	9.6	32.7	15,982,378
North Carolina	4.6	12.9	8,049,313
Texas	4.6	9.3	20,851,820
New York	3.6	12.5	18,976,457
Tennessee	3.4	9.7	5,689,283

7. Estimate the probability of being injured by a lightning strike in New York.
   A 0.0000007%
   B 0.0000002%
   C 0.00007%
   D 0.000002%
8. Estimate the probability of being killed by lightning in North Carolina.
   F 0.0000006%
   G 0.00006%
   H 0.00002%
   J 0.000002%
9. Estimate the probability of being struck by lightning in Florida.
   A 0.00006%
   B 0.00026%
   C 0.0000026%
   D 0.0006%
10. In which two states do you have the highest probability of being struck by lightning?
    F Florida, North Carolina
    G Florida, Tennessee
    H Texas, New York
    J North Carolina, Tennessee

LESSON 10-2

Reading Strategies

Make Predictions

Experimental probability is a statement of the results of a number of trials.

$$\text{Probability} = \frac{\text{number of times an event happens}}{\text{total number of trials}}$$

When you spin this spinner, it could land on the section with dots, the striped section, or the white section.

Use the spinner to answer the following questions.

1. Predict which section the spinner will land on most often. Why?
   the striped section, because it is the largest section
2. Predict which section the spinner will land on least often. Why?
   the dotted section, because it is the smallest section

The actual outcome of an experiment may or may not match your predictions. This chart shows the outcomes of 500 trials.

Outcome	Striped	Spotted	White
Spins	163	152	185

Answer the following questions.

3. How many times did the spinner land on the striped section? 163
4. Did P(striped) match your prediction for the outcome of the experiment? Explain.
   No; The striped section is the largest, but the spinner landed more times on the white section.
5. Did P(spotted) match your prediction for the outcome of the experiment? Explain.
   Yes; The spotted section is the smallest and the spinner landed on it the least number of times.

LESSON 10-2

Puzzles, Twisters & Teasers

Probable Problems

Estimate the probability of drawing a yellow marble in each situation below. Use the letters to answer the riddle.

1. probability: 20.1 % E

Outcome	Blue	Yellow	Green	Black
Draws	217	201	295	287

2. probability: 13 % V

Outcome	Blue	Yellow	Green	White
Draws	33	13	46	8

3. probability: 17 % C

Outcome	Blue	Yellow	Green	White	Red
Draws	20	17	32	18	13

4. probability: 18 % H

Outcome	Blue	Red	Yellow	White	Green
Draws	30	18	18	21	13

5. probability: 11 % D

Outcome	Blue	Yellow	Green	Black	White
Draws	17	11	25	28	19

6. probability: 3.5 % A

Outcome	Blue	Yellow	Green	White	Red	Black
Draws	65	35	235	180	369	116

7. probability: 21 % N

Outcome	Blue	Yellow	Green	Black	White	Brown
Draws	7	21	15	18	19	20

8. probability: 28.5 % W

Outcome	Blue	Red	Yellow
Draws	448	267	285

9. probability: 23.3 % Y

Outcome	Blue	Yellow	Green
Draws	330	233	437

10. probability: 35 % L

Outcome	Blue	Yellow
Draws	65	35

What do Christmas and a cat on the beach have in common?

They both H (18) A (3.5) V (13) E (20.1) SA N (21) D (11) Y (23.3)

C (17) L (35) A W (28.5) S

LESSON 10-3

Practice A

Use a Simulation

1. At a local salad bar, 5 out of every 7 customers order Cobb salad. What is the probability to the nearest percent that a customer will order a Cobb salad? 71%

Use the table of random numbers to simulate the situation for Exercises 2–4.

7	6	3	6	4	5	2	2	4	3
4	5	4	7	5	3	7	4	4	5
6	4	7	4	4	6	4	7	5	2
3	7	1	3	3	3	3	5	7	1
1	2	6	2	5	5	4	4	2	5
4	3	5	2	3	3	7	4	4	3
4	4	5	1	6	2	4	6	7	5
5	4	5	6	6	3	6	3	2	3
3	6	3	2	7	2	2	4	4	4
5	7	5	2	4	5	2	4	7	6

2. Let the numbers 1–5 represent people who order Cobb salad. Checking in rows, how many people would you have to survey before you find 5 people who ordered Cobb salad? 8

3. What is the probability represented by Exercise 2? 62.5%

4. Of the 100 customers represented in the random number table, what is the probability to the nearest percent that the customer will order a Cobb salad? 76%

5. The owner of the salad bar makes the Cobb salad the special of the week. This increases sales to 6 out of 7 customers ordering a Cobb salad. What is the probability to the nearest percent that a customer will order the special? 86%

6. Use the numbers 1–6 in the random number table to represent the customers who purchased a Cobb salad. Of the 100 customers represented in the random number, what is the probability to the nearest percent that the customer will order the special Cobb salad? 88%

LESSON 10-3

Practice B

Use a Simulation

Use the table of random numbers for the problems below.

8125	4764	7693	3675	1642	7988	7048	9135	3138	3256
9566	4413	7215	7992	4320	7438	3805	5473	8847	2397
7336	5393	8623	8570	5095	5685	6695	3570	3605	4656
6470	6065	8239	2953	5942	6496	8899	0701	5368	2106
5210	2570	8137	3587	3578	6657	6636	7188	5717	1770
4329	4110	2655	8258	9928	3873	5609	3695	7091	0368
5315	2654	0484	4601	4336	6624	5403	5870	8545	3905
2361	9097	3753	2498	0544	0923	6099	1737	4025	1221
2677	7741	5342	9844	3722	5120	8742	1382	2842	7386
3292	5084	1130	2747	0664	9718	6072	9432	7008	2024

Mr. Domino gave the same math test to all three of his math classes. In the first two classes, 80% of the students passed the test. If the third class has 20 students, estimate the number of students who will pass the test.

1. Using the first row as the first trial, count the successful outcomes and name the unsuccessful outcomes.

   16 out of 20 successful; 81, 93, 88, 91

2. Count and name the successful outcomes in the second row as the second trial.

   16 out of 20 successful; 95, 92, 88, 97

Determine the successful outcomes in the remaining rows of the random number table.

3. third row 14
4. fourth row 16
5. fifth row 17
6. sixth row 16
7. seventh row 18
8. eighth row 16
9. ninth row 16
10. tenth row 16

11. Based on the simulation, estimate the probability that 80% of the class will pass the math test. 90%

LESSON 10-3

Practice C

Use a Simulation

Use the table of random numbers to simulate each situation. Use at least 10 trials for each simulation. Possible answers are given.

27768	56420	77775	39422	60423	71178	54012	21367
20182	54386	85157	89029	26369	14161	82065	86070
36558	13616	68098	21724	29916	78974	29433	52156
79405	19383	84186	06775	48080	31018	91551	25107
29426	00966	47941	68043	93813	86586	59854	01309
89215	46632	30988	79412	64601	22042	71379	05616
49880	60994	09374	10377	54878	80433	05994	58575
07468	27779	94664	39250	48561	54763	07733	73850
13742	93176	26563	62102	55681	16113	97148	24914
59149	94667	11891	63282	07489	04578	48465	82794

1. A survey of students in the eighth grade shows that 72% of them are wearing or have worn braces. Estimate the probability that 7 out of 10 eighth grade students wear braces or have worn braces. 60%

2. A driving school advertises that 88% of those taking their course pass their driving test. Estimate the probability that 9 out of 10 people who take the school's training will pass their driving test. 60%

3. An ice cream store stocks it shelves with 20% black raspberry chip ice cream because 20% of their customers choose that as their favorite flavor. Estimate the probability that 2 out of every 10 customers of the ice cream store will purchase black raspberry chip ice cream. 50%

4. In the first semester of the year, 61% of the class has been absent at least one day. Estimate the probability that 6 out of 10 students will be absent at least one day in the second semester. 50%

5. 42% of the students surveyed in the eighth grade have more than one sibling. Estimate the probability that 4 out of 10 students in the rest of the school have more than one sibling. 60%

LESSON 10-3 Reteach
Use a Simulation

Situation: Strout's Market is having a contest. They give a puzzle piece to each customer at the checkout. A customer who collects all 10 different puzzle pieces gets $100 in store credit.

Using a table of random numbers, you can model the situation to estimate how many times a customer would have to shop to collect all 10 puzzle pieces.

- Start anywhere in the table. Count the numbers you pass as you "collect" the digits 0-9.

3	1	9	4	1	1	8	8
5	7	4	5	7	7	9	0
7	0	3	0	1	3	5	0
0	4	3	8	9	5	3	8
2	6	1	7	6	7	6	9
0	8	2	6	5	5	9	2

Suppose you start at the top of Column 3 and move to the right. List each number until you have collected all the numbers 0–9.

9 4 1 1 8 8 5 7 4 5 7 7 9 0 7 0 3 0 1 3 5 0 0 4 3 8 9 5 3 8 2 6

You had to go through 32 numbers to get each number at least once (underscored).

- Do the experiment again.

Suppose you start at the bottom of Column 4 and move to the right. When you reach the end of the row, go to the beginning of the table.

6 5 5 9 2 3 1 9 4 1 1 8 8 5 7 4 5 7 7 9 0

You had to go through 21 numbers to get each number at least once (underscored).

- Find the average of your results. $\frac{32 + 21}{2} = \frac{53}{2} = 26.5$

So, on average, you need to shop 27 times to get all 10 pieces to win $100 credit.

Model each situation. Use the list of random numbers shown above. Do two trials. Tell where you start for each trial. Possible answers shown.

1. A box of Whammos contains a toy dinosaur. If there are 10 different model dinosaurs in the collection, estimate how many boxes of Whammos you would have to buy to get all 10 dinosaurs.

   top Col. 2, go right, 33; bottom Col. 5, go right, 38; on average, about 36 boxes

2. For this spinner, estimate how many times you would have to spin the pointer to get the numbers 1–10.

   top Col.8, next row, 27; bottom Col. 1, right, 18; about 23 times

LESSON 10-3 Challenge
Rolling and Tossing

To design a simulation, you may use different devices, such as number cubes or coins.

Situation: At Sonia's Spa, two-thirds of the female clients come to lose weight. For an article about spas, a female client at Sonia's was interviewed. What is the probability that this woman is at the spa to lose weight?

Simulation: To model a ratio of $\frac{2}{3}$, you can use a number cube so that

4 of the outcomes–1, 2, 3, 4–represent *came to lose weight and* 2 of the outcomes–5, 6–represent *did not come to lose weight.*

Then, $P(\text{came to lose weight}) = \frac{4}{6}$, or $\frac{2}{3}$.

To carry out this simulation, Kim rolled a number cube 10 times, with the following results: 4 5 2 6 6 6 1 2 4 3

1. How many of the 10 trials resulted in a woman who came to the spa to lose weight? **6**

2. Find *P*(came to lose weight). Answer as a ratio and as a percent. **$\frac{3}{5}$ or 60%**

Situation: A study shows that a new medication has a 50% chance of curing the condition for which it is prescribed. Keith's doctor prescribes the medication for him. What is the probability that the medication will cure Keith's condition?

3. Using a cube numbered 1–6, describe a simulation.

   Possible answer: 1, 2, 3 represent *cures*; 4, 5, 6 represent *does not cure.*

4. Carry out your simulation for 10 trials. Calculate *P*(cures). Answer as a ratio and as a percent.

   Answers will vary.

5. Using a coin, describe a simulation.

   Possible answer: heads represents *cures*; tails represents *does not cure.*

6. Carry out your simulation for 10 trials. Calculate *P*(cures). Answer as a ratio and as a percent.

   Answers will vary.

LESSON 10-3 Problem Solving
Use a Simulation

Use the table of random numbers below. Use at least 10 trials to simulate each situation. Write the correct answer.

87244	11632	85815	61766
19579	28186	18533	24633
74581	65633	54238	32848
87549	85976	13355	46498
53736	21616	86318	77291
24794	31119	48193	44869
86585	27919	65264	93557
94425	13325	16635	25840
18394	73266	67899	38783
94228	23426	76679	41256

1. Of people 18–24 years of age, 49% do volunteer work. If 10 people ages 18–24 were chosen at random, estimate the probability that at least 4 of them do volunteer work.

   Possible answer: 80%

2. In the 2000 Presidential election, 56% of the population of North Carolina voted for George W. Bush. If 10 people were chosen at random from North Carolina, estimate the probability that at least 8 of them voted for Bush.

   Possible answer: 10%

3. Forty percent of households with televisions watched the 2001 Super Bowl game. If 10 households with televisions are chosen at random, estimate the probability that at least 3 watched the 2001 Super Bowl.

   Possible answer: 90%

Use the table above and at least 10 trials to simulate each situation. Choose the letter for the best estimate.

4. As of August 2000, 42% of U.S. households had Internet access. If 10 households are chosen at random, estimate the probability that at least 5 of them will have Internet access.

   A 0%   C 60%
   (B) 30%   D 90%

5. On average, there is rain 20% of the days in April in Orlando, FL. Estimate the probability that it will rain at least once during your 7-day vacation in Orlando in April.

   F 20%   (H) 70%
   G 50%   J 40%

6. Kareem Abdul-Jabaar is the NBA lifetime leader in field goals. During his career, he made 56% of the field goals he attempted. In a given game, estimate the probability that he would make at least 6 out of 10 field goals.

   (A) 40%   C 80%
   B 60%   D 100%

7. At the University of Virginia 39% of the applicants are accepted. If 10 applicants to the University of Virginia are chosen at random, estimate the probability that at least 4 of them are accepted to the University of Virginia.

   F 10%   H 80%
   (G) 40%   J 70%

LESSON 10-3 Reading Strategies
Analyze Information

A **simulation** is an experiment that models a real situation.

You can simulate flipping a coin without actually using a coin.

Random numbers can simulate flipping a coin. *Random* means happening by chance. No outcome is more likely than another.

You could set up an experiment using random numbers as follows:

- An even number shows an outcome of heads.
- An odd number shows an outcome of tails.

Here are five numbers that have occurred randomly:

9 6 9 1 6

1. Is 9 an even or odd number? What outcome does 9 represent?

   odd; tails

2. Is 6 an even or odd number? What outcome does 6 represent?

   even; heads

3. Use H for heads and T for tails. List the outcomes for the five random numbers.

   T, H, T, T, H

4. From the five outcomes, what is *P*(heads)? $\frac{2}{5}$

5. From the five outcomes, what is the *P*(tails)? $\frac{3}{5}$

This table of ten random numbers simulates flipping a coin. An even number represents heads. An odd number represents tails.

4 2 6 3 3 7 4 6 8 1

Use the random number table to answer the following questions.

8. Use H for heads and T for tails. List the outcomes in the table.

   H, H, H, T, T, T, H, H, H, T

9. From these outcomes, what is *P*(heads)? $\frac{6}{10}$

10. From these outcomes, what is *P*(tails)? $\frac{4}{10}$

LESSON 10-3
Puzzles, Twisters & Teasers
No Cheating!

Circle the words from the list that you find. Find a word that answers the riddle. Circle it and write it on the line.

simulation random numbers model set
table digit probability reasonable strategy

```
S I M U L A T I O N A R T
E T D I G I T E T U S E Y
T A R A N D O M P M D A U
G I T A B L E G H B F S I
N J I P T T G B U E G O P
F T C H E E T A H R H N O
B J I K O P G D E S J A L
M O D E L D U Y O P K B K
P R O B A B I L I T Y L V
Q W E R T Y U I O P H E B
```

What large animal cheats on a test? A cheetah.

LESSON 10-4
Practice A
Theoretical Probability

An experiment consists of tossing two coins.

1. List all the possible outcomes. HH; HT; TH; TT
2. What is the probability of tossing a head and a tail? $\frac{1}{2}$
3. What is the probability of the outcomes being the same? $\frac{1}{2}$

An experiment consists of rolling a fair number cube. Find the probability of each event.

4. $P(6)$ $\frac{1}{6}$
5. $P(1)$ $\frac{1}{6}$
6. $P(\text{odd number})$ $\frac{1}{2}$
7. $P(>4)$ $\frac{1}{3}$

Find the probability of each event using two number cubes.

8. $P(\text{rolling two 5s})$ $\frac{1}{36}$
9. $P(\text{total shown} = 4)$ $\frac{1}{12}$
10. $P(\text{total shown} = 2)$ $\frac{1}{36}$
11. $P(\text{total shown} < 4)$ $\frac{1}{12}$
12. $P(\text{total shown} > 11)$ $\frac{1}{36}$
13. $P(\text{rolling two even numbers})$ $\frac{1}{4}$
14. A bag contains 9 red marbles and 4 blue marbles. How many clear marbles should be added to the bag so the probability of drawing a red marble is $\frac{3}{5}$? 2 clear marbles
15. In a game two fair number cubes are rolled. To make the first move, you need to roll an even total. What is the probability of rolling an even total? $\frac{1}{2}$

LESSON 10-4
Practice B
Theoretical Probability

An experiment consists of rolling one fair number cube. Find the probability of each event.

1. $P(3)$ $\frac{1}{6}$
2. $P(7)$ 0
3. $P(1 \text{ or } 4)$ $\frac{1}{3}$
4. $P(\text{not } 5)$ $\frac{5}{6}$
5. $P(< 5)$ $\frac{2}{3}$
6. $P(> 4)$ $\frac{1}{3}$
7. $P(2 \text{ or odd})$ $\frac{2}{3}$
8. $P(\leq 3)$ $\frac{1}{2}$

An experiment consists of rolling two fair number cubes. Find the probability of each event.

9. $P(\text{total shown} = 3)$ $\frac{1}{18}$
10. $P(\text{total shown} = 7)$ $\frac{1}{6}$
11. $P(\text{total shown} = 9)$ $\frac{1}{9}$
12. $P(\text{total shown} = 2)$ $\frac{1}{36}$
13. $P(\text{total shown} = 4)$ $\frac{1}{12}$
14. $P(\text{total shown} = 13)$ 0
15. $P(\text{total shown} > 8)$ $\frac{5}{18}$
16. $P(\text{total shown} \leq 12)$ 1
17. $P(\text{total shown} < 7)$ $\frac{5}{12}$
18. A bag contains 9 pennies, 8 nickels, and 5 dimes. How many quarters should be added to the bag so the probability of drawing a dime is $\frac{1}{6}$? 8 quarters
19. In a game two fair number cubes are rolled. To make the first move, you need to roll a total of 6, 7, or 8. What is the probability that you will be able to make the first move? $\frac{4}{9}$

LESSON 10-4
Practice C
Theoretical Probability

An experiment consists of rolling two fair number cubes. Find the probability of each event.

1. $P(\text{total shown} = 5)$ $\frac{1}{9}$
2. $P(\text{total shown} > 3)$ $\frac{11}{12}$
3. $P(\text{total shown} > 10)$ $\frac{1}{12}$
4. $P(\text{total shown} < 12)$ $\frac{35}{36}$
5. $P(\text{total shown} \geq 7)$ $\frac{7}{12}$
6. $P(\text{total shown} \leq 4)$ $\frac{1}{6}$

Three separate jars each contain 2 different color marbles. Jar A has a red and a blue marble. Jar B has a red and a green marble. Jar C has a purple and a white marble. One marble is drawn from each jar. The table shows a sample space with all outcomes equally likely. Find each probability.

Jar A	Jar B	Jar C	Outcome
R	R	P	RRP
R	R	W	RRW
R	G	P	RGP
R	G	W	RGW
B	R	P	BRP
B	R	W	BRW
B	G	P	BGP
B	G	W	BGW

7. $P(\text{RRP})$ $\frac{1}{8}$
8. $P(\text{BGW})$ $\frac{1}{8}$
9. $P(\text{2 red with another color})$ $\frac{1}{4}$
10. $P(\text{a green with two other colors})$ $\frac{1}{2}$
11. $P(\text{1 white or 1 purple})$ 1

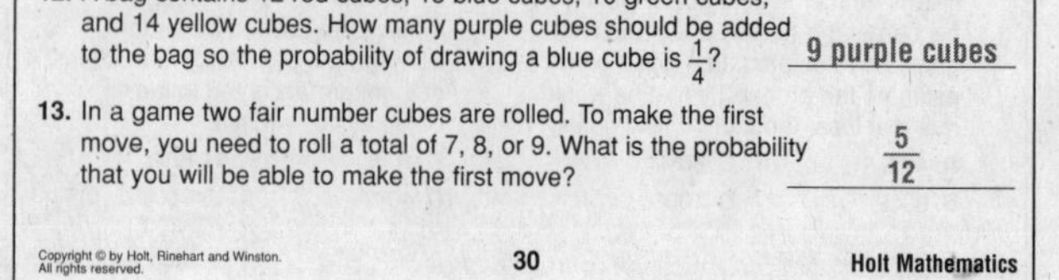

12. A bag contains 12 red cubes, 15 blue cubes, 10 green cubes, and 14 yellow cubes. How many purple cubes should be added to the bag so the probability of drawing a blue cube is $\frac{1}{4}$? 9 purple cubes
13. In a game two fair number cubes are rolled. To make the first move, you need to roll a total of 7, 8, or 9. What is the probability that you will be able to make the first move? $\frac{5}{12}$

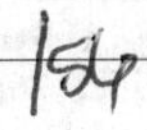

LESSON **10-4**

Reteach

Theoretical Probability

The sample space for a fair coin has 2 possible outcomes: heads or tails. Both possibilities have the same chance of occurring; they are **equally likely.**

The probability of each outcome is $\frac{1}{2}$.
$P(\text{heads}) = P(\text{tails}) = \frac{1}{2}$

Complete to find each probability.

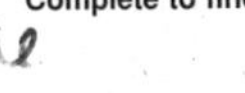

1. 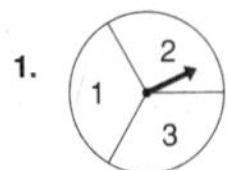

2. 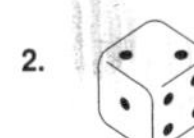 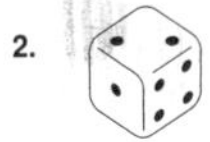

	1.	2.
How many outcomes in the sample space?	3	6
Are the outcomes equally likely?	yes	yes
What is the probability for each outcome?	$\frac{1}{3}$	$\frac{1}{6}$

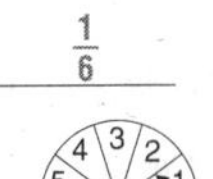

For this spinner, there are 10 possible outcomes in the sample space. The outcomes are equally likely.

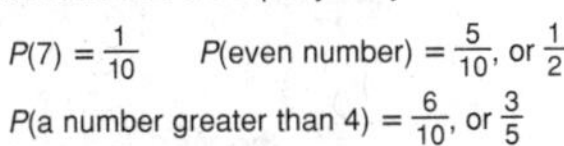

$P(7) = \frac{1}{10}$ $\quad P(\text{even number}) = \frac{5}{10}$, or $\frac{1}{2}$

$P(\text{a number greater than 4}) = \frac{6}{10}$, or $\frac{3}{5}$

When the possible outcomes are equally likely, you calculate the probability that an event E will occur by using a ratio.

$$P(E) = \frac{\text{number of favorable outcomes}}{\text{total number of possible outcomes}}$$

Find each probability.

3. 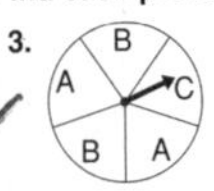

4. 

5.

$P(C) = \frac{1}{5}$

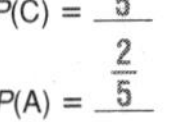

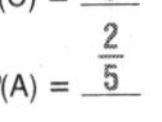

$P(A) = \frac{2}{5}$

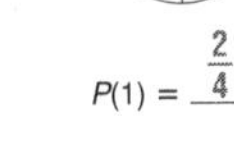

$P(1) = \frac{2}{4}$, or $\frac{1}{2}$

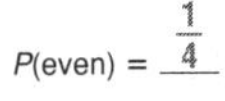

$P(\text{even}) = \frac{1}{4}$

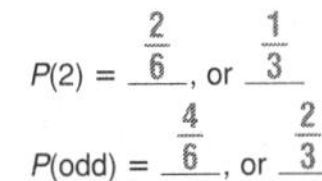

$P(2) = \frac{2}{6}$, or $\frac{1}{3}$

$P(\text{odd}) = \frac{4}{6}$, or $\frac{2}{3}$

LESSON **10-4**

Reteach

Theoretical Probability (continued)

For this spinner:

$P(\text{odd}) = \frac{3}{6}$, or $\frac{1}{2}$ $\quad P(\text{even}) = \frac{1}{6}$

You cannot get an odd number and an even number in the same spin. $\quad P(\text{odd } and \text{ even}) = 0$

Events that cannot occur in the same trial are called **mutually exclusive.**

A number is drawn from {−6, −4, 0, 2, 4, 7, 9}. List the possible favorable results for each event. Tell if the events are mutually exclusive.

6. *Event A*: get an odd number
7, 9
Event B: get a negative number
−6, −4
Are *A* and *B* mutually exclusive? Explain.
yes; no number in both

7. *Event C*: get a multiple of 3
−6, 9
Event D: get an even number
−6, −4, 0, 2, 4
Are *C* and *D* mutually exclusive? Explain.
no; −6 in both events

For the spinner at the top of this page:
$P(\text{odd}) = \frac{3}{6}$, or $\frac{1}{2}$ $\quad P(\text{even}) = \frac{1}{6}$ $\quad P(\text{odd } or \text{ even}) = \frac{3}{6} + \frac{1}{6} = \frac{4}{6}$, or $\frac{2}{3}$

A number is drawn from {−6, −4, 0, 5, 6, 7, 9}. Find the indicated probabilities.

8. odd numbers are: 5, 7, 9
numbers < 0 are: −6, −4
$P(\text{odd}) = \frac{3}{7}$
$P(\text{number} < 0) = \frac{2}{7}$
$P(\text{odd number or number} < 0) =$
$\frac{3}{7} + \frac{2}{7} = \frac{5}{7}$

9. numbers > 6 are: 7, 9
even numbers: −6, −4, 0, 6
$P(\text{number} > 6) = \frac{2}{7}$
$P(\text{even number}) = \frac{4}{7}$
$P(\text{number} > 6 \text{ or even number}) =$
$\frac{2}{7} + \frac{4}{7} = \frac{6}{7}$

LESSON **10-4**

Challenge

Picture This

Venn diagrams can be used to illustrate and solve problem situations involving probability.

Consider a cube numbered 1–6.

Let *Event A* = rolling an even number on the cube.
favorable outcomes = 2, 4, 6

Let *Event B* = rolling a number less than 5 on the cube.
favorable outcomes = 1, 2, 3, 4

Note that the numbers 2 and 4 are in both events and, thus, lie in the intersection of the two circles that represent *Events A* and *B.*

So, to determine the probability of getting an even number that is also less than 5, the favorable outcomes are in the intersection of the circles.

$$P(A \text{ and } B) = \frac{\text{number of favorable outcomes}}{\text{total number of possible outcomes}} = \frac{2}{6}, \text{ or } \frac{1}{3}.$$

Then, to determine the probability of getting an even number *or* a number that is less than 5, count the elements in the intersection only once.

$$P(A \text{ or } B) = \frac{\text{number of favorable outcomes}}{\text{total number of possible outcomes}} = \frac{5}{6}$$

Draw a Venn diagram to solve each problem. A cube numbered 1–6 is rolled once.

1. Find the probability of getting an odd number that is greater than 2.
Event A = a number that is odd
Event B = a number that is >2

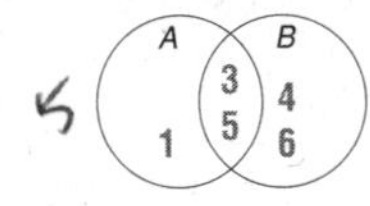

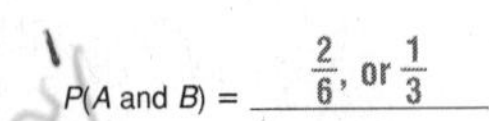

$P(A \text{ and } B) = \frac{2}{6}$, or $\frac{1}{3}$

2. Find the probability of getting an even number or a number less than 3.
Event A = a number that is even
Event B = a number that is <3

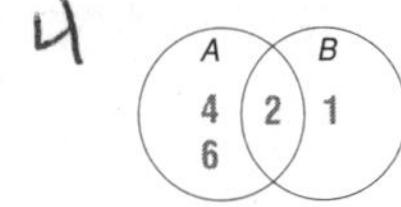

$P(A \text{ or } B) = \frac{4}{6}$, or $\frac{2}{3}$

LESSON **10-4**

Problem Solving

Theoretical Probability

A company that sells frozen pizzas is running a promotional special. Out of the next 100,000 boxes of pizza produced, randomly chosen boxes will be prize winners. There will be one grand prize winner who will receive $100,000. Five hundred first prize winners will get $1000, and 3,000 second prize winners will get a free pizza. Write the correct answer in fraction and percent form.

1. What is the probability that the box of pizza you just bought will be a grand prize winner?
$\frac{1}{100,000}$; 0.001%

2. What is the probability that the box of pizza you just bought will be a first prize winner?
$\frac{1}{200}$; 0.5%

3. What is the probability that the box of pizza you just bought will be a second prize winner?
$\frac{3}{100}$; 3%

4. What is the probability that you will win anything with the box of pizza you just bought?
$\frac{3,501}{100,000}$; 3.501%

Researchers at the National Institutes of Health are recommending that instead of screening all people for certain diseases, they can use a Punnett square to identify the people who are most likely to have the disease. By only screening these people, the cost of screening will be less. Fill in the Punnett square below and use them to choose the letter for the best answer.

	D	d
D	DD	Dd
d	Dd	dd

5. What is the probability of DD?
A 0%
B 25% (circled)
C 50%
D 75%

6. What is the probability of Dd?
F 25%
G 50% (circled)
H 75%
J 100%

7. What is the probability of dd?
A 0%
B 25% (circled)
C 50%
D 75%

8. DD or Dd indicates that the patient will have the disease. What is the probability that the patient will have the disease?
F 25%
G 50%
H 75% (circled)
J 100%

LESSON **10-4**

Reading Strategies
Draw Conclusions

Theoretical probability describes what might be expected to happen in an event. It helps you draw conclusions.

Probability(event) = $\frac{\text{number of ways an event can occur}}{\text{total number of events}}$

When you flip a coin, there are two possible events.

→ land on heads

→ land on tails

You have a 1 out of 2 chance for each.

$P(\text{heads}) = \frac{1}{2}$ $P(\text{tails}) = \frac{1}{2}$

These events have the same probability, so you can draw the conclusion that they are **equally likely** to occur.

A number cube has six faces. The numbers on the six faces are 2, 3, 4, 2, 3, and 4. Answer the following about this number cube.

1. How many ways can the number cube land? 6 ways
2. How many ways can you get a 2 on the cube? 2 ways
3. What is *P*(rolling a 2)? $\frac{2}{6}$ or $\frac{1}{3}$
4. How many ways can you get an even number? 4 ways

5. What is *P*(rolling an even number)? $\frac{4}{6}$ or $\frac{2}{3}$

6. Can you conclude that *P*(rolling a 2) and *P*(rolling an even number) are equally likely to occur? Explain.

No: The probability of rolling a 2 is $\frac{2}{6}$ and the probability of rolling of an even number is $\frac{4}{6}$.

7. What can you conclude about rolling an even number or an odd number?

You are more likely to roll an even number than an odd number.

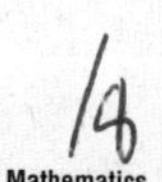

LESSON **10-4**

Puzzles, Twisters & Teasers
That's Odd!

Using the chart, write the probability for each outcome as a fraction. Unscramble the letters to answer the riddle. The fractions under the riddle will give you hints to get you started.

HTT Probability: $\frac{1}{4}$ E

THT Probability: $\frac{1}{8}$ O

TTT Probability: $\frac{1}{8}$ P

HHT Probability: $\frac{1}{4}$ F

HHH Probability: $\frac{1}{8}$ A

TTH Probability: $\frac{1}{4}$ S

HTH Probability: $\frac{1}{8}$ R

THH Probability: $\frac{1}{4}$ K

TH Probability: 0 C

THHT Probability: 0 Y

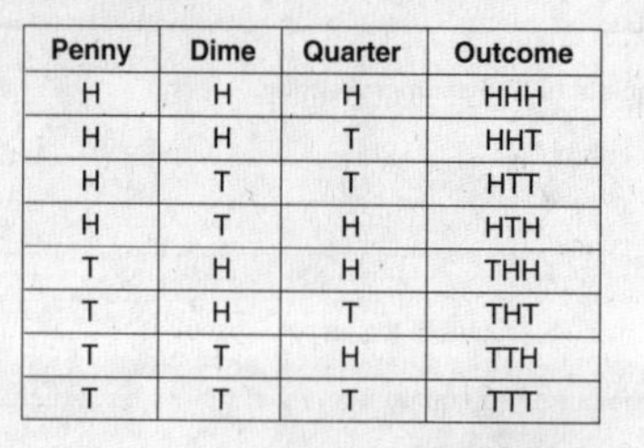

Penny	Dime	Quarter	Outcome
H	H	H	HHH
H	H	T	HHT
H	T	T	HTT
H	T	H	HTH
T	H	H	THH
T	H	T	THT
T	T	H	TTH
T	T	T	TTT

What did the boy do when he wanted to see time fly?

He dropped his clock

O ($\frac{1}{8}$) F ($\frac{1}{4}$) F A

S ($\frac{1}{4}$) K ($\frac{1}{4}$) Y (0) S C (0) R ($\frac{1}{8}$) A ($\frac{1}{8}$) P ($\frac{1}{8}$) E ($\frac{1}{4}$) R

Name ________________ Date __________ Class __________

LESSON **10-5**

Practice A
Independent and Dependent Events

Determine if the events are dependent or independent.

1. drawing a card from a deck of cards and tossing a coin

independent

2. drawing two cards from a regular deck of cards and not replacing the first

dependent

3. spinning a number on a spinner and drawing a marble from a container

independent

4. drawing two red marbles without replacement from a container of red and blue marbles

dependent

An experiment consists of spinning each spinner once. Find the probability. For each spin, all outcomes are equally likely.

5. *P*(A and 2) $\frac{1}{12} \approx 0.08\overline{3}$

6. *P*(D and 1) $\frac{1}{8} = 0.125$

7. *P*(B and 3)

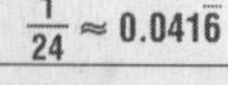

8. *P*(B and 1 or 3) $\frac{1}{6} \approx 0.1\overline{6}$

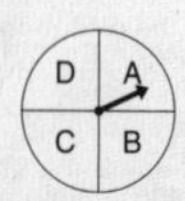

9. *P*(C and 1 or 2)

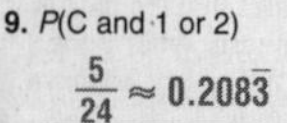

10. *P*(A and not 1)

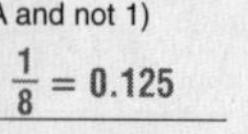

11. Georgiana wants to toss three coins and get all heads. What is the probability of tossing 3 coins and getting 3 heads?

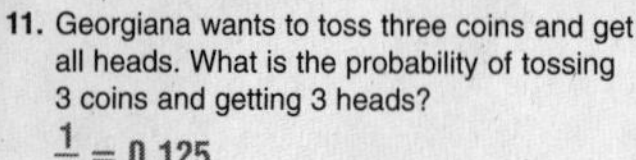

LESSON **10-5**

Practice B
Independent and Dependent Events

Determine if the events are dependent or independent.

1. choosing a tie and shirt from the closet independent
2. choosing a month and tossing a coin independent
3. rolling two fair number cubes once, then rolling them again if you received the same number on both number cubes on the first roll dependent

An experiment consists of rolling a fair number cube and tossing a fair coin.

4. Find the probability of getting a 5 on the number cube and tails on the dime. $\frac{1}{12}$
5. Find the probability of getting an even number on the number cube and heads on the dime. $\frac{1}{4}$
6. Find the probability of getting a 2 or 3 on the number cube and heads on the dime. $\frac{1}{6}$

A box contains 3 red marbles, 6 blue marbles, and 1 white marble. The marbles are selected at random, one at a time, and are not replaced. Find the probability.

7. *P*(blue and red) $\frac{1}{5} = 0.2$
8. *P*(white and blue) $\frac{1}{15} \approx 0.0\overline{66}$
9. *P*(red and white) $\frac{1}{30} \approx 0.0\overline{33}$
10. *P*(red and white and blue) $\frac{1}{40} = 0.025$
11. *P*(red and red and blue) $\frac{1}{20} = 0.05$
12. *P*(red and blue and red) $\frac{1}{8} = 0.125$
13. *P*(red and red and red) $\frac{1}{120} = 0.00\overline{833}$
14. *P*(white and blue and blue) $\frac{1}{24} = 0.041\overline{6}$
15. *P*(white and red and white) 0

LESSON **10-5**

Practice C

Independent and Dependent Events

Consider a regular deck of cards without the jokers. Cards are replaced after each draw. Find the probability of each of the following.

1. P(pair of red kings)

   $\frac{1}{676} \approx 0.00148$

2. P(a diamond and a black seven)

   $\frac{1}{104} \approx 0.00962$

Use the same deck of cards but do not replace the card after each draw.

3. P(ace of hearts and king of hearts)

   $\frac{1}{2652} \approx 0.000377$

4. P(a ten and a jack)

   $\frac{4}{663} \approx 0.00603$

5. P(red card and a black card)

   $\frac{13}{51} \approx 0.2549$

6. P(a club and king or red ace)

   $\frac{1}{34} \approx 0.0294$

7. Mr. and Mrs. Reginald are expecting their first baby. The doctor tells them they are having triplets. What is the probability that the babies will all be the same sex?

   $\frac{1}{4} = 0.25$

8. Sid has a bag of 12 red, 14 brown, and 10 blue marbles. He chooses one, shoots it, and chooses another. What is the probability that his first selection is a red marble, and then a blue marble?

   $\frac{2}{21} \approx 0.0952$

9. If Justine's initials are JMD, what is the probability that she will draw her initials from a box containing the letters of the alphabet? There is no replacement of letters after each is drawn.

   $\frac{1}{15,600} \approx 0.000064$

10. There are 13 math students, 10 science students, and 17 English students in a group. If only one prize is allowed per person, what is the probability that the moderator will award a science student a prize and then award another prize to a math student?

    $\frac{1}{12} \approx 0.08\overline{3}$

LESSON **10-5**

Reteach

Independent and Dependent Events

Carlos is to draw 2 straws at random from a box of straws that contains 4 red, 4 white, and 4 striped straws.

P(1st straw is striped) $= \frac{4}{12}$ ← number of striped straws / ← total number of straws

If Carlos *returns* the 1st straw to the box before drawing the 2nd straw, the probability that the 2nd straw is striped remains the same.

P(2nd straw is striped)

$= \frac{4}{12}$ ← same number of striped straws / ← same total number of straws

When the 1st straw is returned before the 2nd draw, the 2nd draw occurs as though the 1st draw never happened, **independent events.**

P(striped and striped) $= \frac{4}{12} \times \frac{4}{12}$

$= \frac{1}{3} \times \frac{1}{3} = \frac{1}{9}$

If Carlos *does not return* the first straw to the box before drawing the second straw, the probability that the second straw is striped changes.

P(2nd straw is striped)

$= \frac{3}{11}$ ← one striped straw has been taken / ← one less straw in total number

When the 1st straw is not returned before the 2nd draw, the number of straws remaining is changed, **dependent events.**

P(striped and striped) $= \frac{4}{12} \times \frac{3}{11}$

$= \frac{1}{3} \times \frac{3}{11} = \frac{1}{11}$

Describe the events as independent or dependent.

1. Josh tosses a coin and spins a spinner. independent
2. Ana draws a colored toothpick from a jar. Without replacing it, she draws a second toothpick. dependent
3. Sue draws a card from a deck of cards and replaces it. Then she draws a second card from the deck. independent

Each situation begins with a box of marbles that contains 2 red, 3 blue, 4 green, and 3 yellow marbles. Complete to find each probability.

4. A 1st marble is drawn and replaced. Then a 2nd marble is drawn.

   P(red and blue) $= \frac{2}{12} \times \frac{3}{12} = \frac{1}{24}$

5. A 1st marble is drawn and not replaced. A 2nd marble is drawn.

   P(red and blue) $= \frac{2}{12} \times \frac{3}{11} = \frac{1}{22}$

6. A 1st marble is drawn and replaced. Then a 2nd marble is drawn.

   P(red and red) $= \frac{2}{12} \times \frac{2}{12} = \frac{1}{36}$

7. A 1st marble is drawn and not replaced. A 2nd marble is drawn.

   P(red and red) $= \frac{2}{12} \times \frac{1}{11} = \frac{1}{66}$

LESSON **10-5**

Challenge

Probability from a Table

The table shows the results of a survey of 50 students. The students were asked whether they liked a newly released movie.

	Yes	No
Male	16	14
Female	12	8

According to the table:

16 + 14, or 30 males were surveyed.
12 + 8, or 20 females were surveyed.
16 + 12, or 28, of those surveyed liked the movie.
14 + 8, or 22, of those surveyed did not like the movie.

The probability that a student randomly selected from the group is a male is $\frac{30}{50}$, or $\frac{3}{5}$.

The probability that a student did not like the movie is $\frac{22}{50}$, or $\frac{11}{25}$.

The probability that a student is a male who did not like the movie is $\frac{14}{50}$, or $\frac{7}{25}$.

The table below shows the results of a survey of 50 students. The students were asked whether they liked a newly released CD. One student was selected at random from this group. Use the table to solve.

	Yes	No
Male	35	5
Female	8	2

1. What is the probability that the selected student liked the CD? $\frac{43}{50}$
2. What is the probability that the student did not like the CD? $\frac{7}{50}$
3. What is the probability that the student is a female? $\frac{1}{5}$
4. What is the probability that the student is a female who liked the movie? $\frac{4}{25}$
5. What is the probability that the student is a male who did not like the movie? $\frac{1}{10}$

LESSON **10-5**

Problem Solving

Independent and Dependent Events

Are the events independent or dependent? Write the correct answer.

1. Selecting a piece of fruit, then choosing a drink.

   Independent events

2. Buying a CD, then going to another store to buy a video tape if you have enough money left.

   Dependent events

Dr. Fred Hoppe of McMaster University claims that the probability of winning a pick 6 number game where six numbers are drawn from the set 1 through 49 is about the same as getting 24 heads in a row when you flip a fair coin.

3. Find the probability of winning the pick 6 game and the probability of getting 24 heads in a row when you flip a fair coin.

   game: $\frac{1}{13,983,816}$

   Coin: $\frac{1}{16,777,216}$

4. Which is more likely: to win a pick 6 game or to get 24 heads in a row when you flip a fair coin?

   Pick 6 game

In a shipment of 20 computers, 3 are defective. Choose the letter for the best answer.

5. Three computers are randomly selected and tested. What is the probability that all three are defective if the first and second ones are not replaced after being tested?

   A $\frac{1}{760}$ C $\frac{27}{8000}$
   (B) $\frac{1}{1140}$ D $\frac{3}{5000}$

6. Three computers are randomly selected and tested. What is the probability that all three are defective if the first and second ones are replaced after being tested?

   F $\frac{1}{760}$ (H) $\frac{27}{8000}$
   G $\frac{1}{1140}$ J $\frac{3}{5000}$

7. Three computers are randomly selected and tested. What is the probability that none are defective if the first and second ones are not replaced after being tested?

   (A) $\frac{34}{57}$ C $\frac{4913}{6840}$
   B $\frac{4913}{8000}$ D $\frac{1}{2000}$

8. Three computers are randomly selected and tested. What is the probability that none are defective if the first and second ones are replaced after being tested?

   F $\frac{34}{57}$ H $\frac{4913}{6840}$
   (G) $\frac{4913}{8000}$ J $\frac{1}{2000}$

LESSON 10-5

Reading Strategies

Use Context

When the outcome of one event does not affect the outcome of another, they are called **independent events.**

A cube has the numbers 1–6 on the faces.

P(rolling a 6) is $\frac{1}{6}$.

If you roll the cube a second time, P(rolling a 6) is still $\frac{1}{6}$.

The second roll of the cube is independent of the first roll.

When the outcome of one event affects the probability of another, they are called **dependent events.**

There are 10 cubes in a bag. Three of them are red.

P(drawing a red cube) is $\frac{3}{10}$.

If a red cube is drawn and not replaced, the probability of drawing a red cube on the second draw is 2 out of 9.

So P(drawing another red cube) is $\frac{2}{9}$.

Drawing a cube the second time is dependent on drawing the first cube.

Tell if each situation describes dependent events or independent events.

1. You flip a coin. Then you flip a coin a second time.

   independent events

2. You roll a number cube three times.

   independent events

3. You draw a card from a deck of cards and do not replace the card. Then you draw a second time.

   dependent events

4. You spin a spinner once. Then you spin the spinner a second time.

   independent events

5. You pull a marble out of a jar and leave it on the table. Then you pull another marble out of the jar.

   dependent events

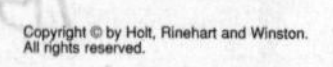

LESSON 10-5

Puzzles, Twisters & Teasers

Declare Your Independence!

Decide whether the events are dependent or independent. Circle the letter above your answer. Unscramble the letters to answer the riddle.

1. tossing a coin twice
   Q dependent — (I) independent
2. pulling two socks from a drawer at the same time
   (E) dependent — P independent
3. drawing two marbles out of a bag
   (B) dependent — D independent
4. spinning a spinner five times
   M dependent — (C) independent
5. spinning two different spinners two times
   K dependent — (E) independent
6. drawing names out of a hat
   (R) dependent — B independent
7. throwing a pair of number cubes ten times
   X dependent — (U) independent
8. throwing one number cube five times
   U dependent — (S) independent
9. picking cards from a deck
   (G) dependent — V independent
10. throwing three coins three times
    A dependent — (R) independent

What do penguins eat for lunch?

I C E B U R G E R S

LESSON 10-6

Practice A

Making Decisions and Predictions

The zoo store sells caps with different animals pictured on the cap. The table shows the animals pictured on the last 100 caps sold. The manager plans to order 1500 new caps.

Animal Caps Sold

Animal	Number
Tiger	30
Orangutan	20
Panda Bear	25
Giraffe	18
Gazelle	7

1. Find the probability of selling a tiger cap.

   $\frac{3}{10}$

2. How many tiger caps should the manager order?

   450 tiger caps

3. Find the probability of selling a panda bear cap.

   $\frac{1}{4}$

4. How many panda bear caps should the manager order?

   375 panda bear caps

5. Use probability to decide how many orangutan caps the manager should order.

   300 orangutan caps

6. Nancy spins the spinner at the right 60 times. Predict how many times the spinner will land on the number 2.

   15 times

Decide whether the game is fair.

7. Roll two fair number cubes labeled 1–6. Player A wins if both numbers are odd. Player B wins if both numbers are even.

   fair: $\frac{1}{4} = \frac{1}{4}$

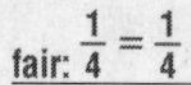

LESSON 10-6

Practice B

Making Decisions and Predictions

A sports store sells water bottles in different colors. The table shows the colors of the last 200 water bottles sold. The manager plans to order 1800 new water bottles.

Water Bottles Sold

Color	Number
Red	30
Blue	50
Green	25
Yellow	10
Purple	10
Clear	75

1. How many red water bottles should the manager order? 270

2. How many green water bottles should the manager order? 225

3. How many clear water bottles should the manager order? 675

4. If the carnival spinner lands on 10, the player gets a large stuffed animal. Suppose the spinner is spun 30 times. Predict how many large stuffed animals will be given away. 5

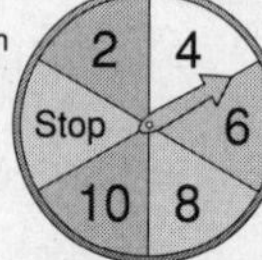

Decide whether the game is fair.

5. Roll two fair number cubes labeled 1–6. Player A wins if both numbers are the same. Player B wins if both numbers are different.

   not fair: $\frac{1}{6} \neq \frac{5}{6}$

6. Roll two fair number cubes labeled 1–6. Add the numbers. Player A wins if the sum is 5 or less. Player B wins if the sum is 9 or more.

   fair: $\frac{5}{18} = \frac{5}{18}$

7. Toss three fair coins. Player A wins if exactly one tail lands up. Otherwise, Player B wins.

   not fair: $\frac{3}{8} \neq \frac{5}{8}$

LESSON **10-6**

Practice C

Making Decisions and Predictions

A fair number cube is labeled 1–6. Predict the number of outcomes for the given number of rolls.

1. outcome: 5
   number of rolls: 42
   7 times
2. outcome: less than 3
   number of rolls: 75
   25 times
3. outcome: not 4
   number of rolls: 60
   50 times
4. outcome: 2, 3, or 4
   number of rolls: 30
   15 times
5. outcome: greater than 2
   number of rolls: 48
   32 times
6. outcome: multiple of 3
   number of rolls: 90
   30 times
7. In his last eight 5K runs, Jeremy had the following times in minutes: 24:48, 23:45, 23:12, 24:08, 25:36, 22:03, 23:29, and 24:01. Based on these results, what is the best prediction of the number of times Jeremy will run faster than 24 minutes in his next 20 5K runs?
   10 times
8. Before a school vote on a mascot for a community river project, a sample of students surveyed gave the otter 18 votes, the osprey 8 votes, and the beaver 6 votes. Based on these results, predict the number of votes for each animal if 1200 students vote.
   otter: 675 votes; osprey: 300 votes; beaver: 225 votes

Decide whether each game is fair.

9. A spinner is divided evenly into 6 sections. There are 3 green sections, 2 blue sections, and 1 yellow section. Player A wins if the spinner does not land on green. Otherwise, Player B wins.
   fair: $\frac{1}{2} = \frac{1}{2}$
10. Roll two fair number cubes labeled 1–6. Add the numbers. Player A wins if the sum is 8 or more. Player B wins if the sum is 5 or less.
    not fair: $\frac{5}{12} \neq \frac{5}{18}$
11. Toss three fair coins. Player A wins if exactly two tails land up. Player B wins if all heads or all tails land up.
    not fair: $\frac{3}{8} \neq \frac{1}{4}$

LESSON **10-6**

Reteach

Making Decisions and Predictions

Probability can be used to make predictions about data.

The spinner has 5 equal sections. To predict how many times you will land on the number 1 in 30 spins, first find the probability of landing on 1 in one spin. Then multiply 30 spins by that probability.

Step 1 Find the probability.

$$P(1) = \frac{\text{number of 1s}}{\text{number of equal sections}} = \frac{2}{5}$$

Step 2 Multiply the number of spins by the probability.
There are 30 spins and $P(1) = \frac{2}{5}$.

$$30 \times \frac{2}{5} = 6 \times 2 = 12$$

You will spin about 12 1s in 30 spins.

Each situation begins with a box of marbles that contains 2 red, 3 blue, 4 green, and 3 yellow marbles. Complete to find each probability.

1. Predict how many times you will land on C in 32 spins.
   a. $P(C) = \frac{1}{4}$
   b. $32 \times P(C) = 32 \times \frac{1}{4} = 8$
   c. about 8 times

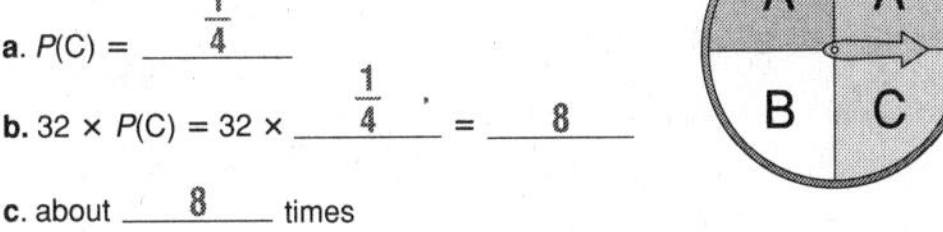

2. Predict how many times you will land on B in 48 spins.
   a. $P(B) = \frac{1}{4}$
   b. $48 \times P(B) = 48 \times \frac{1}{4} = 12$
   c. about 12 times
3. Predict how many times you will land on A in 50 spins.
   a. $P(A) = \frac{1}{2}$
   b. $50 \times \frac{1}{2} = 25$
   c. about 25 times

LESSON **10-6**

Challenge

Complimentary Events

The complement of an event A is the event that A does not occur. It includes all outcomes not in A. Represent the complement of *A* as not *A*.

Since either *A* or not *A* must occur, $P(A) + P(\text{not } A) = 1$. This means that $P(A) = 1 - P(\text{not } A)$ and $P(\text{not } A) = 1 - P(A)$.

Suppose *A* is the event rolling 4 when rolling a fair number cube. Then the complement of *A* is the event rolling a number that is not 4.

$$P(4) = \frac{1}{6} \qquad P(\text{not } 4) = 1 - \frac{1}{6} = \frac{5}{6}$$

Two fair number cubes are rolled. Solve.

1. Suppose *A* is the event rolling a sum of 2 or 3.
   a. Find $P(A) = \frac{1}{12}$
   b. Use words to describe the complement of *A*.
   Possible answer: rolling a sum greater than 3
   c. Find $P(\text{not } A)$ $\frac{11}{12}$
2. Suppose *A* is the event rolling a sum of at least 10.
   a. $P(A) = \frac{1}{6}$
   b. Use words to describe the complement of *A*.
   Possible answer: rolling a sum less than 10
   c. Find $P(\text{not } A)$. $\frac{5}{6}$
3. Suppose *A* is the event rolling a sum of 8.
   a. $P(A) = \frac{5}{36}$
   b. Use words to describe the complement of *A*.
   Possible answer: rolling a sum other than 8
   c. Find $P(\text{not } A)$. $\frac{31}{36}$

LESSON **10-6**

Problem Solving

Making Decisions and Predictions

Write the correct answer.

1. A quality control inspector at a light bulb factory finds 2 defective bulbs in a batch of 1000 light bulbs. If the plant manufactures 75,000 light bulbs in one day, predict how many will be defective.
   150 defective bulbs
2. A game consists of rolling two fair number cubes labeled 1–6. Add both numbers. Player A wins if the sum is greater than 10. Player B wins if the sum is 7. Is the game fair or not? Explain.
   not fair: $\frac{1}{12} \neq \frac{1}{6}$
3. A spinner has 5 equal sections numbered 1–5. Predict how many times Kevin will spin an even number in 40 spins.
   16 times
4. In her last six 100-meter runs, Lee had the following times in seconds: 12:04, 13:11, 12:25, 11:58, 12:37, and 13:20. Based on these results, what is the best prediction of the number of times Lee will run faster than 13 seconds in her next 30 runs?
   20 times

Use the table below that shows the number of colors of the last 200 T-shirts sold at a T-shirt shop. The manager of the store wants to order 1800 new T-shirts. Choose the letter of the best answer

T-Shirts Sold

Color	Number
Red	35
Blue	55
Green	15
Black	65
White	30

5. How many red T-shirts should the manager order?
   A 175　C 378
   (B) 315　D 630
6. How many blue T-shirts should the manager order?
   (F) 495　H 900
   G 665　J 990
7. How many more black T-shirts than white T-shirts should the manager order?
   A 855　(C) 315
   B 585　D 270

LESSON 10-6 Reading Strategies
Draw Conclusions

You can use probability to make decisions about data.

The table shows the colors of 100 wrist bands sold at a sports shop.

Wrist Bands Sold

Color	Number
Yellow	40
Blue	20
Pink	30
Red	10

The store manager wants to order 2000 new bands. She can use the data in the table to decide how many of each color of wrist band to order in the next order. The manager should order the greatest number of yellow bands and the fewest number of red bands.

To find the number of red bands, first use the data in the table to find *P*(*red*), the probability of selling a red band.

$P(red) = \frac{10}{100} = 0.1$

Then use the probability to find how many red bands to order:

0.1 of 2000 = 0.1 × 2000 = 200

So, the manager should order 200 red bands.

Answer each question. Use the information above.

1. What is the probability of selling a yellow band?

0.4

2. Write an expression that represents how many yellow bands the manager should order.

0.4 × 2,000

3. How many yellow bands should the manager order?

800

4. What is the probability of selling a pink band?

0.3

5. How many pink bands should the manager order?

600

LESSON 10-6 Puzzles, Twisters & Teasers
A Friendly Solution!

Find and circle words from the list in the word search horizontally, vertically, or diagonally. Find a word that answers the riddle. Circle it and write it on the line.

probability decision prediction fair unfair
game proportion chance likely favorable

What kind of ship never sinks? **Friendship**

LESSON 10-7 Practice A
Odds

1. Complete the table by finding the odds in favor and the odds against an event based on the probability of the event.

Probability	$\frac{1}{4}$	$\frac{2}{5}$	$\frac{1}{8}$	$\frac{3}{4}$	$\frac{3}{7}$
Odds in favor	1:3	2:3	1:7	3:1	3:4
Odds against	3:1	3:2	7:1	1:3	4:3

2. Complete the table with the missing information.

Probability	$\frac{3}{8}$	$\frac{2}{3}$	$\frac{1}{8}$	$\frac{8}{9}$	$\frac{3}{10}$
Odds in favor	3:5	2:1	1:7	8:1	3:7
Odds against	5:3	1:2	7:1	1:8	7:3

If there are 28 boys and 22 girls in the music class, find the odds for each of the following.

3. What are the odds in favor of selecting a boy as the conductor?

28:22 = 14:11

4. What are the odds in favor of selecting a girl as the conductor?

22:28 = 11:14

5. What are the odds against selecting a boy as the conductor?

22:28 = 11:14

6. What are the odds against selecting a girl as the conductor?

28:22 = 14:11

7. If the probability of drawing a red card from a regular deck of cards without the jokers is $\frac{1}{2}$ what are the odds in favor of and against drawing a red card?

50:50 or 1:1

LESSON 10-7 Practice B
Odds

A bag contains 9 red marbles, 5 green marbles, and 6 purple marbles.

1. Find *P*(red marble)

$\frac{9}{20} = 0.45$

2. Find *P*(green marble)

$\frac{1}{4} = 0.25$

3. Find *P*(purple marble)

$\frac{3}{10} = 0.3$

4. Find the odds in favor of choosing a red marble.

9:11

5. Find the odds against choosing a red marble.

11:9

6. Find the odds in favor of choosing a green marble.

5:15 = 1:3

7. Find the odds against choosing a green marble.

15:5 = 3:1

8. Find the odds in favor of choosing a purple marble.

6:14 = 3:7

9. Find the odds against choosing a purple marble.

14:6 = 7:3

10. Find the odds in favor of not choosing a green marble.

15:5 = 3:1

11. Find the odds in favor of choosing a red or purple marble.

15:5 = 3:1

12. If the probability of Helena winning the contest is $\frac{2}{5}$, what are the odds in favor of Helena winning the contest?

2:3

13. The odds in favor of the Bruins winning the Stanley Cup are 5 to 4. What is the probability that the Bruins will win the Stanley Cup?

$\frac{5}{9} \approx 0.55\overline{5}$

LESSON 10-7

Practice C
Odds

Use the spinner to find the following odds. The spinner turns but the pointer stays in one place.

1. Find the odds in favor of the spinner stopping at 1.

   1:7

2. Find the odds against the spinner stopping at 5.

   7:1

3. Find the odds in favor of the spinner stopping at 2.

   2:6 = 1:3

4. Find the odds against the spinner stopping at white.

   4:4 = 1:1

5. Find the odds in favor of the spinner stopping at dots or 1.

   3:5

6. Find the odds of the spinner stopping at lines and an even number.

   2:6 = 1:3

Katera won a contest in math class. As her prize she could pick one envelope from 18 different envelopes. The prizes included a pass to the local amusement park, 6 movie passes, a gift certificate for a school hat, 8 free lunch passes, and 2 gift certificates to the school supply store. Each envelope contained one prize.

Find the odds in favor of each of the following.

7. Katera choosing the envelope containing the amusement park pass.

   1:17

8. Katera choosing an envelope containing a free lunch pass.

   8:10 = 4:5

9. Katera choosing an envelope containing a gift certificate for the school supply store.

   2:16 = 1:8

10. Katera choosing an envelope containing a movie pass.

   6:12 = 1:2

LESSON 10-7

Reteach
Odds

Baseball fans do not usually ask "What is the probability that the New York Yankees will win the World Series this year?"

Fans who want to know the chances of a team winning usually ask "What are the *odds* that the Yankees will win?"

Odds that an event *E* will or will not occur can be defined as a ratio of probabilities.

odds in favor $= \frac{P(E)}{P(\text{not } E)}$ **odds against** $= \frac{P(\text{not } E)}{P(E)}$

What are the odds in favor of getting a 4 in one roll of a numbered cube?

$P(4) = \frac{1}{6}$ $P(\text{not } 4) = \frac{5}{6}$ $\text{odds}(4) = \frac{P(4)}{P(\text{not } 4)} = \frac{\frac{1}{6}}{\frac{5}{6}} = \frac{1}{5}$

So, the odds in favor of getting a 4 are 1 to 5.

Complete to find the indicated odds. In each case, a cube numbered 1–6 is rolled once.

1. Find the odds in favor of getting a number greater than 4.

   $P(> 4) = \frac{2}{6}$ $P(\text{not} > 4) = \frac{4}{6}$ $\text{odds}(> 4) = \frac{P(> 4)}{P(\text{not} > 4)} = \frac{\frac{2}{6}}{\frac{4}{6}} = \frac{2}{4}$, or $\frac{1}{2}$

   So, the odds in favor of getting a number greater than 4 are 1 to 2.

2. Find the odds against getting a 3.

   $P(3) = \frac{1}{6}$ $P(\text{not } 3) = \frac{5}{6}$ $\text{odds}(\text{not } 3) = \frac{P(\text{not } 3)}{P(3)} = \frac{\frac{5}{6}}{\frac{1}{6}} = \frac{5}{1}$

   So, the odds against getting a 3 are 5 to 1.

3. Find the odds in favor of getting an even number.

   $P(\text{even}) = \frac{3}{6}$ $P(\text{not even}) = \frac{3}{6}$ $\text{odds}(\text{even}) = \frac{P(\text{even})}{P(\text{not even})} = \frac{\frac{3}{6}}{\frac{3}{6}} = \frac{3}{3}$, or $\frac{1}{1}$

   So, the odds in favor of getting an even number are 1 to 1.

LESSON 10-7

Reteach
Odds, continued

Convert each probability ratio to an odds ratio.

4. The probability of winning a prize is $P(\text{win}) = \frac{2}{25}$.

   $P(\text{not win}) = 1 - \frac{2}{25}$, or $\frac{23}{25}$ $\text{odds}(\text{win}) = \frac{P(\text{win})}{P(\text{not win})} = \frac{\frac{2}{25}}{\frac{23}{25}}$ or $\frac{2}{23}$, or 2 to 23

5. The probability of not winning a prize $P(\text{not win}) = \frac{100}{109}$.

   $P(\text{win}) = 1 - \frac{100}{109}$, or $\frac{9}{109}$ $\text{odds}(\text{not win}) = \frac{P(\text{not win})}{P(\text{win})} = \frac{\frac{100}{109}}{\frac{9}{109}}$ or $\frac{100}{9}$, or 100 to 9

Reconsider a problem in finding odds by using probabilities.

What are the odds in favor of getting a 4 in one roll of a numbered cube?

$P(4) = \frac{1}{6}$ $P(\text{not } 4) = \frac{5}{6}$ $\text{odds}(4) = \frac{P(4)}{P(\text{not } 4)} = \frac{\frac{1}{6}}{\frac{5}{6}} = \frac{1}{5}$

Study the ratio for odds with respect to its relation to the probability ratios.

1 ← numerator of probability in favor
5 ← numerator of probability against
1 + 5 = 6 ← denominator of probability ratio

If the odds in favor of an event are $\frac{a}{b}$, or $a{:}b$, then the probability that the event will occur is $\frac{a}{a + b}$.

If the odds in favor of being chosen for a committee are 2:3, then the probability of being chosen is $\frac{2}{2 + 3}$, or $\frac{2}{5}$.

Convert each odds ratio to a probability ratio.

6. The odds in favor of winning a prize are 3:20.

   The probability of winning a prize is: $\frac{3}{3 + 20}$, or $\frac{3}{23}$.

7. The odds against winning a prize are 100:7.

   The probability of not winning a prize is: $\frac{100}{100 + 7}$, or $\frac{100}{107}$.

LESSON 10-7

Challenge
All Sizes and Shapes

The most common shaped die is a cube numbered 1 through 6. However, dice come in a variety of shapes. The illustrations show a cube and five other polyhedral dice. All but one of these six dice are regular polyhedrons.

The table below shows the probability of rolling a 1 and the odds in favor of rolling a 1, not a 1, an even number, and a number that is a multiple of 5.

	A	B	C	D	E	F	G
1	Number of sides	Name of shape	*P*(1)	Odds(1)	Odds (not 1)	Odds (even)	Odds (multiple of 5)
2	4	tetrahedron	0.25	1:3	3:1	1:1	0:4
3	6	cube	0.16	1:5	5:1	1:1	1:5
4	8	octahedron	0.125	1:7	7:1	1:1	1:7
5	10	decahedron	0.1	1:9	9:1	1:1	2:8
6	12	dodecahedron	0.083	1:11	11:1	1:1	1:5

Answer each question.

1. The formula for cell C2 is $\frac{1}{A2}$. What does the A2 represent in the probability formula?

   total number of outcomes

2. The first number in the ratio in cell F5 is $\frac{A5}{2}$. What does this number represent?

   How many even sides there are on the die

3. Complete the rest of the table. See chart above.

LESSON 10-7
Problem Solving
Odds

In the last 25 Summer Olympics since 1900, an American man has won the gold medal in the 400-meter dash 18 times. Write the correct answer.

1. Find the probability that an American man will win the gold medal in the 400-meter dash in the next Summer Olympics.

   $\frac{18}{25}$

2. Find the probability that an American man will not win the gold medal in the 400-meter dash in the next Summer Olympics.

   $\frac{7}{25}$

3. Find the odds that an American man will win the gold medal in the 400-meter dash in the next Summer Olympics.

   18:7

4. Find the odds that an American man will not win the gold medal in the 400-meter dash in the next Summer Olympics.

   7:18

Use the table below that shows the probability that a player will end up on a certain square after a single roll in a game of Monopoly.

Probability of Ending Up on a Monopoly Square

Square	Probability	Rank
In Jail	$\frac{39}{1000}$	1
Illinois Ave.	$\frac{32}{1000}$	2
Go	$\frac{31}{1000}$	3
Boardwalk	$\frac{26}{1000}$	18
Park Place	$\frac{22}{1000}$	33

5. What are the odds that you will end up in jail on your next roll in a game of Monopoly?
   A 39:1000
   Ⓑ 39:961
   C 1000:961
   D 961:39

6. What are the odds that you will end up on Boardwalk on your next roll in a game of Monopoly?
   A 13:500 Ⓒ 13:487
   B 500:13 D 487:13

7. What are the odds that you will not end up on Boardwalk on your next roll in a game of Monopoly?
   F 487:500 H 13:487
   G 500:487 Ⓙ 487:13

8. What are the odds that you will end up on Go on your next roll in a game of Monopoly?
   Ⓐ 31:969 C 31:1000
   B 969:31 D 1000:31

9. What are the odds that you will not end up on Park Place on your next roll in a game of Monopoly?
   F 11:489 H 489:500
   Ⓖ 489:11 J 500:489

LESSON 10-7
Reading Strategies
Use a Graphic Organizer

This graphic organizer will help you understand odds.

Definition of Odds	Ways to Write Odds
A ratio between favorable outcomes and unfavorable outcomes	Favorable to unfavorable Favorable:unfavorable Unfavorable to favorable Unfavorable:favorable

Odds

Odds in Favor	Odds Against
A ratio of favorable outcomes to unfavorable outcomes. Of the 35 people who entered the contest, 12 could win a free CD. The odds in favor of winning a CD are 12 to 23.	A ratio of unfavorable outcomes to favorable outcomes. Of the 35 people who entered the contest, 23 could not win a free CD. The odds against winning a CD are 23 to 12.

Use the information in the graphic organizer to help you answer the questions.

1. What is the meaning of "odds"?

   a ratio between favorable outcomes and unfavorable outcomes

2. What ratio is used to describe odds in favor?

   favorable outcomes to unfavorable outcomes

3. What ratio is used to describe odds against?

   unfavorable outcomes to favorable outcomes

4. What are the odds in favor of winning a CD?

   12:23

5. What are the odds against winning a CD?

   23:12

LESSON 10-7
Puzzles, Twisters & Teasers
Even Odder!

Solve the crossword puzzle. Use the formula $a{:}b$ = odds in favor.

Across

2. The odds in ___ of an event is the ratio of favorable to unfavorable outcomes.
4. Odds and ___ are not the same thing.
5. a = number of ___ outcomes
9. If the probability of an event is $\frac{1}{3}$, this means that on average it will happen in one out of every three ___.

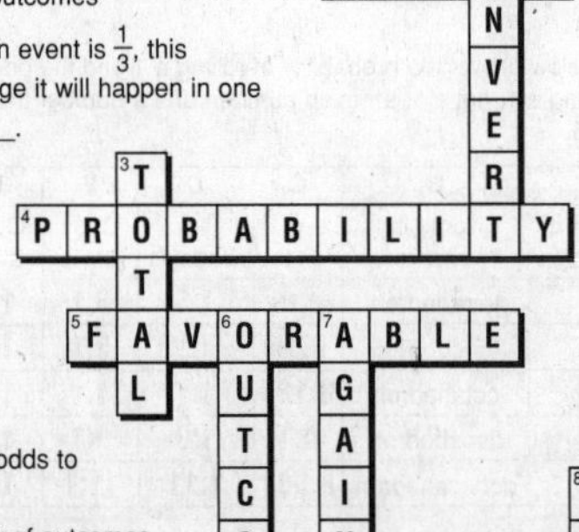

Down

1. It is possible to ___ odds to probabilities.
3. $a + b$ = ___ number of outcomes
6. b = number of unfavorable ___
7. The odds ___ an event is the ratio of unfavorable outcomes to favorable outcomes.
8. If the ___ in favor of an event are $a{:}b$, then the probability of the event occurring is $\frac{a}{a+b}$.

LESSON 10-8
Practice A
Counting Principles

1. A snack bar serves tea, juice, and milk in small, medium, and large sizes. List all the different possible beverage orders.

   small tea, small juice, small milk;
   medium tea, medium juice, medium milk;
   large tea, large juice, large milk

2. The school's football team has a choice of different colored jerseys and different colored pants to wear for their uniforms. They have purple jerseys, white jerseys, and striped jerseys. The choices for the pants are purple or white. List all the different possible uniforms the team can wear.

   purple jerseys with purple pants
   purple jerseys with white pants
   white jerseys with purple pants
   white jerseys with white pants
   striped jerseys with purple pants
   striped jerseys with white pants

3. What is the probability that the team will select the white jerseys with purple pants?

   $\frac{1}{6}$

4. Student identification codes at a high school are 4-digit randomly generated codes beginning with 1 letter and ending with 3 numbers. If all codes are equally likely, how many possible codes are there?

   26,000

5. Find the probability of being assigned the code A123.

   $\frac{1}{26{,}000}$

6. Fabiana bought 3 fashion magazines, 2 exercise magazines, and 2 dance magazines. How many choices of magazines does she have to read?

   7

LESSON **10-8**
Practice B
Counting Principles

Employee identification codes at a company contain 2 letters followed by 2 numbers. All codes are equally likely.

1. Find the number of possible identification codes.

   67,600

2. Find the probability of being assigned the code MT49.

   $\frac{1}{67,600} \approx 0.000015$

3. Find the probability that an ID code of the company does not contain the letter *A* as the second letter of the code.

   $\frac{65,000}{67,600} = \frac{25}{26} \approx 0.962$

4. Find the probability that an ID code of the company does not contain the number 2.

   $\frac{54,756}{67,600} = \frac{81}{100} = 0.81$

5. Mrs. Sharpe is planning her dinners for next week. The choices for the entree are roast beef, turkey, or pork. The choices of carbohydrates are mashed potatoes, baked potatoes, or noodles. The vegetable choices are broccoli, spinach, or carrots. Make a tree diagram indicating the possible outcomes for each entree.

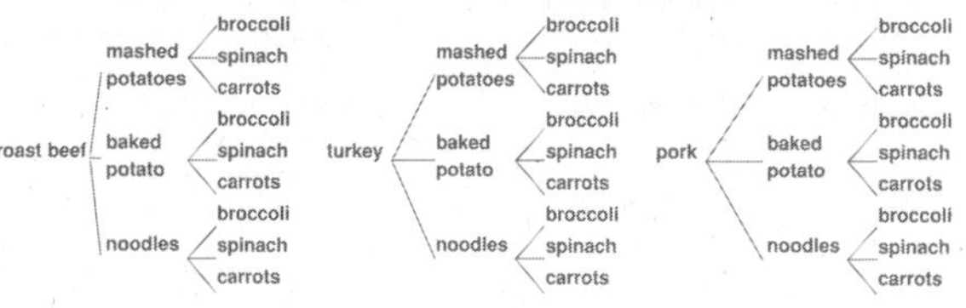

6. How many different meals could Mrs. Sharpe prepare? 27

Find the probability for each of the following.

7. *P*(dinner with baked potato)

   $\frac{1}{3} = 0.\overline{333}$

8. *P*(dinner with noodles and carrots)

   $\frac{1}{9} = 0.\overline{111}$

9. Mitch bought 2 sports magazines, 3 guitar magazines, and 3 news magazines. How many choices of magazines does he have to read?

   8

LESSON **10-8**
Practice C
Counting Principles

Find the number of possible outcomes.

1. pasta: spaghetti, linguine

   sauce: pesto, Alfredo, marinara

   6

2. music: country, pop, rap

   artist: male, female, duo, group

   12

3. eye color: blue, brown, green

   hair color: black, blond, brown, red

   sex: male, female

   24

4. font: Arial, Calligraphy, Helvetica

   size: 10, 12, 14, 16, 20, 22, 24

   color: black, red, blue, green

   84

5. sport: baseball, basketball, football, hockey, soccer, volleyball

   level: professional, college, high school, grade school 24

Use the chart for Exercises 6 and 7.

Main Color	Trim Color	Frame Size	Tire Size	Gears
blue	white	19 in.	24 in.	15 speed
green	black	21 in.	26 in.	24 speed
red	yellow	23 in.		

6. Janis plans to buy a bike. How many combinations are possible with a choice of one main color, one trim color, one frame size, one tire size, and one gear selection? 108

7. Janis decides to buy a green bike. How many combinations are now possible? 36

A computer randomly generates a 5-character password of 3 letters followed by 2 digits. All passwords are equally likely.

8. Find the probability that a password contains exactly one 2. $\frac{9}{50} = 0.18$

9. Find the probability that a password contains exactly one *A*. $\frac{1875}{17,576} \approx 0.1067$

LESSON **10-8**
Reteach
Counting Principles

The Fundamental Counting Principle can help you solve some problems about situations that involve more than one activity.

the number of ways in which one activity can be performed × **the number of ways in which a second activity can be performed** = **the total number of ways in which both activities can be performed**

Apply the Fundamental Counting Principle to find the total number of possibilities in each situation.

1. Kelly has 6 shirts and 4 coordinating pants. The number of possible shirt-pants outfits is: 6 × 4, or 24

2. The menu for dinner lists 2 soups, 4 meats, and 3 desserts. How many different meals that have one soup, one meat, and one dessert are possible? 2 × 4 × 3, or 24

A **tree diagram** helps you see all the possibilities in a sample space.

If three coins are tossed at the same time, list all the possible outcomes.

List, in a column, the 2 possibilities for the 1st coin.

For each possibility for the 1st coin, list the 2 possibilities for the 2nd coin.

For each possibility for the 2nd coin, list the 2 possibilities for the 3rd coin.

Read the diagram across to write the list of all possible outcomes.

In this situation, there are 2 × 2 × 2 = 8 possible outcomes.

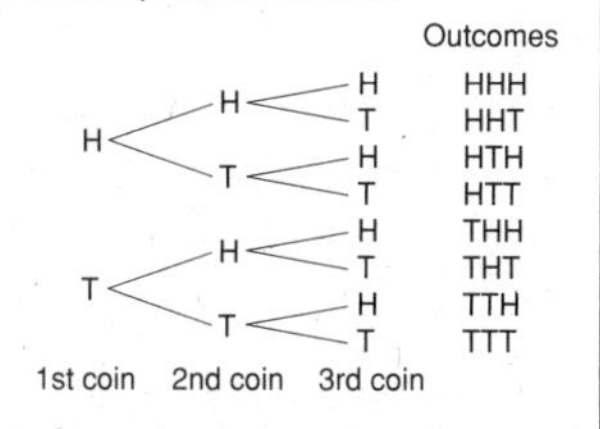

Draw a tree diagram and list the outcomes.

3. A vendor is selling cups of ice cream. There are 2 different sizes of cups: small (S), or large (L). There are 2 different flavors of ice cream: vanilla (V) or chocolate (C). There are 2 different toppings: fudge (F) or pineapple (P).

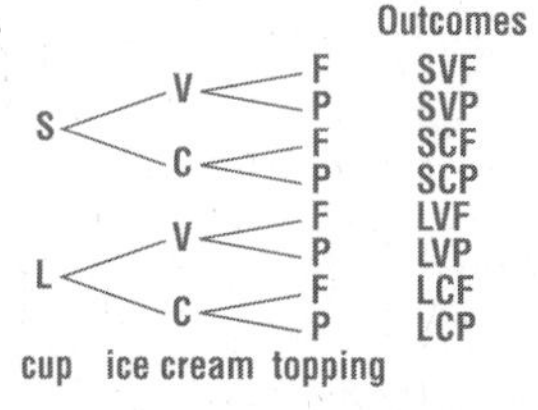

LESSON **10-8**
Reteach
Counting Principles (continued)

How many different 5-letter "words" are possible using the letters of TRIANGLE? Letters can be used only once in each "word."

There are 8 choices for the 1st letter, 7 choices for the 2nd letter, 6 choices for the 3rd letter, 5 choices for the 4th letter, and 4 choices for the 5th.

Apply the Fundamental Counting Principle.

$\frac{8}{\text{1st letter}} \times \frac{7}{\text{2nd letter}} \times \frac{6}{\text{3rd letter}} \times \frac{5}{\text{4th letter}} \times \frac{4}{\text{5th letter}} = 6720$ possibilities

If a "word" is selected at random from the 6720 possibilities, what is the probability that it will be the "word" ANGLE?

There is only one outcome ANGLE. $P(\text{ANGLE}) = \frac{1}{6720}$

If a "word" is selected at random from the 6720 possibilities, what is the probability that it will not contain the letter G?

Find the number of favorable outcomes.

Eliminate the letter G from the choices. So, there are 7 choices to begin.

$\frac{7}{\text{1st letter}} \times \frac{6}{\text{2nd letter}} \times \frac{5}{\text{3rd letter}} \times \frac{4}{\text{4th letter}} \times \frac{3}{\text{5th letter}} = 2520$ possibilities

$P(\text{5-letter "word" with no G}) = \frac{\text{number of favorable outcomes}}{\text{total number of possible outcomes}} = \frac{2520}{6720}$, or $\frac{3}{8}$

Apply the Fundamental Counting Principle.

4. Consider the letters of the word MEDIAN.

   a. How many different 4-letter "words" are possible? Letters can be used only once.

   $\frac{6}{\text{1st letter}} \times \frac{5}{\text{2nd letter}} \times \frac{4}{\text{3rd letter}} \times \frac{3}{\text{4th letter}} = 360$ possibilities

   b. If a 4-letter "word" is selected at random from all the possibilities, then: $P(\text{DEAN}) = \frac{1}{360}$

   c. If a 4-letter "word" is selected at random from all the possibilities, what is the probability that it will not contain the letter D?

   favorable outcomes: $\frac{5}{\text{1st letter}} \times \frac{4}{\text{2nd letter}} \times \frac{3}{\text{3rd letter}} \times \frac{2}{\text{4th letter}} = 120$

   $P(\text{4-letter "word" with no D}) = \frac{\text{number of favorable outcomes}}{\text{total number of possible outcomes}} = \frac{120}{360}$, or $\frac{1}{3}$

LESSON **10-8**

Challenge

Answer the Phone

The world is divided into 9 telephone numbering zones. The North American Numbering Plan (NANP) was developed in 1947 to enable direct dialing without the need for an operator.

NANP numbers are 10 digits in length, of the form

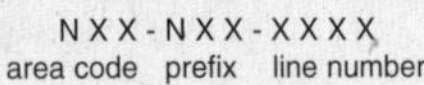

N X X - N X X - X X X X
area code prefix line number

Originally, the plan created 86 areas and allowed for expansion to 144 areas. In 1995, NANP expanded to 792 area codes.

1. For the 3-digit area code NXX, the plan allows N to be any digit 2–9. Currently, there are no restrictions on the other 2 digits of the area code. How many area codes are possible?

8 × 10 × 10, or 800

2. For the 3-digit prefix, the plan allows N to be any digit 2–9. How many line numbers are possible for a given prefix?

10 × 10 × 10 × 10, or 10,000

3. How many telephone numbers are possible for a given area code?

800 × 10,000, or 8,000,000

Some of the prefixes are reserved for services.They are of the form N11 where N is any digit 2–9.

The most familiar service code is 911, reserved for emergency calls. Other currently assigned service codes are 411 (local directory assistance). 611 (repairs), 711 (teletypewriter [hearing/speech impaired]), 811(business office).

4. If all the service code prefixes are removed, how many telephone numbers are possible for a given area code?

8,000,000 − 8 × 10 × 10 × 10 × 10, or 7,920,000

Some other prefixes are not available for general use, such as:

555 (information), 800 and 888 (usually, but not always, toll free), 900 (pay per call).

5. For each prefix that is not available for general use, how many fewer telephone numbers are available for general use?

10,000

LESSON **10-8**

Problem Solving

Counting Principles

Write the correct answer.

1. The 5-digit zip code system for United States mail was implemented in 1963. How many different possibilities of zip codes are there with a 5-digit zip code where each digit can be 0 through 9?

100,000

2. In 1983, the ZIP +4 zip code system was introduced so mail could be more easily sorted by the 5-digit zip code plus an additional 4 digits. How many different possibilities of zip codes are there with the ZIP +4 system?

1,000,000,000

3. In Canada, each postal code has 6 symbols. The first, third and fifth symbols are letters of the alphabet and the second, fourth and sixth symbols are digits from 0 through 9. How many possible postal codes are there in Canada?

17,576,000

4. In the United Kingdom the postal code has 6 symbols. The first, second, fifth and sixth are letters of the alphabet and the third and fourth are digits from 0 through 9. How many possible postal codes are there in the United Kingdom?

45,697,600

Choose the letter for the best answer.

5. In Sharon Springs, Kansas, all of the phone numbers begin 852–4. The only differences in the phone numbers are the last 3 digits. How many possible phone numbers can be assigned using this system?

A 729 C 6561
(B) 1000 D 10,000

6. Many large cities have run out of phone numbers and so a new area code must be introduced. How many different phone numbers are there in a single area code if the first digit can't be zero?

F 90,000 (H) 9,000,000
G 4,782,969 J 10,000,000

7. How many different phone numbers are possible using a 3-digit area code and a 7-digit phone number if the first digit of the area code and phone number cannot be zero?

A 3,486,784,401 C 9,500,000,000
(B) 8,100,000,000 D 10,000,000,000

8. A shipping service offers to send packages by ground delivery using 2 different companies, by next day air using 3 different companies, and by 2-day air using 3 different companies. How many different shipping options does the service offer?

F 3 H 10
(G) 8 J 18

LESSON **10-8**

Reading Strategies

Use a Visual Aid

If you have 2 pairs of shorts and 3 shirts, how many different outfits can you make?

Shorts A

Shorts B

Shirt A

Shirt B

Shirt C

A **tree diagram** is a way to visualize the possible outfits.

2 pairs of shorts → Each paired with one of the 3 shirts → Number of different outfits

Shorts A
- Shirt A → Shorts A & Shirt A
- Shirt B → Shorts A & Shirt B
- Shirt C → Shorts A & Shirt C

Shorts B
- Shirt A → Shorts B & Shirt A
- Shirt B → Shorts B & Shirt B
- Shirt C → Shorts B & Shirt C

Use the tree diagram to answer the following questions.

1. How many different shirts could you pair with Shorts A?

3 different shirts

2. How many different outfits could you make with Shorts A?

3 different outfits

3. How many different shirts could you pair with Shorts B?

3 different shirts

4. How many different outfits can you make with Shorts B?

3 different outfits

5. With 2 different pairs of shorts and 3 different shirts, how many different outfits can you make?

6 different outfits

LESSON **10-8**

Puzzles, Twisters & Teasers

It's FUN-damental!

Determine if the number of possible outcomes is correct for each situation below. Circle the letter next to your answer. Use the letters to solve the riddle.

1. 3 types of birds and 2 types of cages possible outcomes: 6

(I) correct A incorrect

2. 4 colors and 3 sizes possible outcomes: 7

M correct (T) incorrect

3. 3 types of bagels and 3 types of spreads possible outcomes: 9

(J) correct K incorrect

4. 3 destinations and 4 months possible outcomes: 34

P correct (U) incorrect

5. 3 types of soup and 5 types of sandwiches possible outcomes: 15

(S) correct U incorrect

6. 5 destinations and 3 modes of transportation possible outcomes: 8

X correct (W) incorrect

7. 5 shirts, 3 pants, and 2 jackets possible outcomes: 30

(A) correct Z incorrect

8. 4 main dishes, 5 appetizers, and 5 desserts possible outcomes: 14

L correct (V) incorrect

9. 3 types of paint and 1 type of paper possible outcomes: 3

(E) correct B incorrect

10. 10 type fonts, 5 type sizes, and 3 paper sizes possible outcomes: 18

V correct (D) incorrect

What did the wave say to the beach?

Nothing, I T J U S T W A V E D.

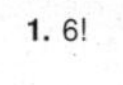

LESSON 10-9

Practice A
Permutations and Combinations

Express each expression as a product of factors.

1. 6! $6 \bullet 5 \bullet 4 \bullet 3 \bullet 2 \bullet 1$

2. 3! $3 \bullet 2 \bullet 1$

3. 7! $7 \bullet 6 \bullet 5 \bullet 4 \bullet 3 \bullet 2 \bullet 1$

4. $\frac{8!}{5!}$ $\frac{8 \bullet 7 \bullet 6 \bullet 5 \bullet 4 \bullet 3 \bullet 2 \bullet 1}{5 \bullet 4 \bullet 3 \bullet 2 \bullet 1}$

5. $\frac{4!}{2!}$ $\frac{4 \bullet 3 \bullet 2 \bullet 1}{2 \bullet 1}$

6. $\frac{9!}{6!}$ $\frac{9 \bullet 8 \bullet 7 \bullet 6 \bullet 5 \bullet 4 \bullet 3 \bullet 2 \bullet 1}{6 \bullet 5 \bullet 4 \bullet 3 \bullet 2 \bullet 1}$

Evaluate each expression.

7. 5! 120
8. 9! 362,880
9. 3! 6
10. 8! 40,320
11. $\frac{7!}{4!}$ 210
12. $\frac{8!}{7!}$ 8
13. $\frac{5!}{2!}$ 60
14. 7! − 5! 4920
15. (6 − 3)! 6
16. $\frac{4!}{(6-2)!}$ 1
17. $\frac{9!}{(8-3)!}$ 3024
18. $\frac{7!}{(9-4)!}$ 42

19. An anagram is a rearrangement of the letters of a word or words to make other words. How many possible arrangements of the letters W, O, R, D, and S can be made? 120

20. Janell is having a group of friends over for dinner and is setting the name cards on the table. She has invited 5 of her friends for dinner. How many different seating arrangements are possible for Janell and her friends at the table? 720

21. How many different selections of 4 books can be made from a bookcase displaying 12 books? 495

LESSON 10-9

Practice B
Permutations and Combinations

Evaluate each expression.

1. 10! 3,628,800
2. 13! 6,227,020,800
3. 11! − 8! 39,876,480
4. 12! − 9! 478,638,720
5. $\frac{15!}{8!}$ 32,432,400
6. $\frac{18!}{12!}$ 13,366,080
7. $\frac{13!}{(17-12)!}$ 51,891,840
8. $\frac{19!}{(15-2)!}$ 19,535,040
9. $\frac{15!}{(18-10)!}$ 32,432,400

10. Signaling is a means of communication through signals or objects. During the time of the American Revolution, the colonists used combinations of a barrel, basket, and a flag placed in different positions atop a post. How many different signals could be sent by using 3 flags, one above the other on a pole, if 8 different flags were available?

336

11. From a class of 25 students, how many different ways can 4 students be selected to serve in a mock trial as the judge, defending attorney, prosecuting attorney, and the defendant?

303,600

12. How many different 4 people committees can be formed from a group of 15 people?

1365

13. The girls' basketball team has 12 players. If the coach chooses 5 girls to play at a time, how many different teams can be formed?

792

14. A photographer has 50 pictures to be placed in an album. How many combinations will the photographer have to choose from if there will be 6 pictures placed on the first page?

15,890,700

LESSON 10-9

Practice C
Permutations and Combinations

Evaluate each expression.

1. $\frac{16!}{(15-4)!}$ 524,160
2. $\frac{21!}{(19-3)!}$ 2,441,880
3. $\frac{17!}{5!(17-5)!}$ 6188
4. ${}_7P_3$ 210
5. ${}_9P_4$ 3024
6. ${}_{10}P_8$ 1,814,400
7. ${}_{18}P_2$ 306
8. ${}_9C_2$ 36
9. ${}_{11}C_5$ 462
10. ${}_{13}C_{11}$ 78
11. ${}_{15}C_3$ 455

12. The music class has 20 students and the teacher wants them to practice in groups of 5. How many different ways can the first group of 5 be chosen?

15,504 groups

13. Math, science, English, history, health, and physical education are the subjects on Jamar's schedule for next year. Each subject is taught in each of the 6 periods of the day. From how many different schedules will Jamar be able to choose?

720 schedules

14. The Hamburger Trolley has 25 different toppings available for their hamburgers. They have a \$3 special that is a hamburger with your choice of 5 different toppings. Assume no toppings are used more than once. How many different choices are available for the special?

53,130 choices

15. Many over the counter stocks are traded through Nasdaq, an acronym for the National Association of Securities Dealers Automatic Quotations. Most of the stocks listed on the Nasdaq use a 4-digit alphabetical code. For example, the code for Microsoft is MSFT. How many different 4-digit alphabetical codes could be available for use by the association? Assume letters cannot be reused.

358,800 codes

LESSON 10-9

Reteach
Permutations and Combinations

Factorial: a string of factors that counts down to 1

$6! = 6 \cdot 5 \cdot 4 \cdot 3 \cdot 2 \cdot 1$

To evaluate an expression with factorials, cancel common factors.

$\frac{5!}{3!} = \frac{5 \cdot 4 \cdot 3 \cdot 2 \cdot 1}{3 \cdot 2 \cdot 1} = 5 \cdot 4 = 20$

Complete to evaluate each expression.

1. $\frac{7!}{4!} = \frac{7 \cdot 6 \cdot 5 \cdot 4 \cdot 3 \cdot 2 \cdot 1}{4 \cdot 3 \cdot 2 \cdot 1}$
$= 7 \cdot$ $6 \cdot 5$ $=$ 210

2. $\frac{6!}{(5-2)!} = \frac{6!}{3!} = \frac{6 \cdot 5 \cdot 4 \cdot 3 \cdot 2 \cdot 1}{3 \cdot 2 \cdot 1}$
$=$ $6 \bullet 5 \bullet 4$ $=$ 120

Permutation: an arrangement in which order is important

wxyz is not the same as *yxzw*

Apply the Fundamental Counting Principle to find how many permutations are possible using all 4 letters *w*, *x*, *y*, *z* with no repetition.

$\frac{4}{\text{1st letter}} \times \frac{3}{\text{2nd letter}} \times \frac{2}{\text{3rd letter}} \times \frac{1}{\text{4th letter}} = 4! = 24$ possible arrangements

When you arrange *n* things, *n*! permutations are possible.

Complete to find the number of permutations.

3. In how many ways can 6 people be seated on a bench that seats 6?

$6! =$ $6 \cdot 5 \cdot 4 \cdot 3 \cdot 2 \cdot 1$

$=$ 720 possibilities

4. How many 5-digit numbers can be made using the digits 7, 4, 2, 1, 8 without repetitions?

$5! =$ $5 \cdot 4 \cdot 3 \cdot 2 \cdot 1$

$=$ 120 possibilities

Apply the Fundamental Counting Principle to find how many permutations are possible using 4 letters 2 at a time, with no repetitions.

$\frac{4}{\text{1st letter}} \times \frac{3}{\text{2nd letter}} = 12$ possible 2-letter arrangements

Apply the Foundamental Counting Principle.

5. In how many ways can 6 people be seated on a bench that seats 4?

$\frac{6}{\text{1st seat}} \times \frac{5}{\text{2nd seat}} \times \frac{4}{\text{3rd seat}} \times \frac{3}{\text{4th seat}} =$ 360 possibilities

6. How many 3-digit numbers can be made using the digits 7, 4, 2, 1, 8 without repetitions?

$\frac{5}{\text{1st digit}} \times \frac{4}{\text{2nd digit}} \times \frac{3}{\text{3rd digit}} =$ 60 possibilities

LESSON 10-9
Reteach
Permutations and Combinations (continued)

When using fewer than the available number of items in an arrangement, instead of the Fundamental Counting Principle, you can use a formula to find the number of possible permutations.

To arrange n things r at a time, the number of possible permutations P is: $_nP_r = \frac{n!}{(n-r)!}$

Find how many permutations are possible using 4 letters 2 at a time, with no repetitions.

$_4P_2 = \frac{4!}{(4-2)!} = \frac{4!}{2!} = \frac{4 \cdot 3 \cdot 2 \cdot 1}{2 \cdot 1} = 12$ possible 2-letter arrangements

Complete to apply the permutations formula.

7. In how many ways can 6 people be seated on a bench that seats 4?

$_6P_4 = \frac{6!}{(6-4)!} = \frac{6!}{2!} = \frac{6 \cdot 5 \cdot 4 \cdot 3 \cdot 2 \cdot 1}{2 \cdot 1} =$ 360 possible seating arrangements

8. How many 3-digit numbers can be made using the digits 7, 4, 2, 1, 8 without repetitions?

$_5P_3 = \frac{5!}{(5-3)!} = \frac{5!}{2!} = \frac{5 \cdot 4 \cdot 3 \cdot 2 \cdot 1}{2 \cdot 1} =$ 60 possible 3-digit numbers

Combination: an arrangement in which order is not important

How many 2-letter combinations can be made from the 4 letters w, x, y, z without repetition?

wx	~~xw~~	~~yw~~	~~zw~~
wy	xy	~~yx~~	~~zx~~
wz	xz	yz	~~zy~~

There are fewer combinations than permutations.

The combinations wx and xw are the same. After all the same combinations are removed, there are 6 different combinations possible.

$_nC_r = \frac{_nP_r}{r!} = \frac{n!}{(n-r)!\,r!}$

The number of combinations C of n things taken r at a time is:

$_4C_2 = \frac{_4P_2}{2!} = \frac{4!}{(4-2)!\,2!} = \frac{4!}{2!\,2!} = \frac{4^2 \cdot 3 \cdot 2 \cdot 1}{2 \cdot 1 \cdot 2 \cdot 1} = 2 \cdot 3 = 6$

Complete to apply the combinations formula.

9. How many different 4-person committees can be formed from a group of 6 people?

$_6C_4 = \frac{_6P_4}{4!} = \frac{6!}{(6-4)!\,4!} = \frac{6!}{2!\,4!} = \frac{6^3 \cdot 5 \cdot 4 \cdot 3 \cdot 2 \cdot 1}{2 \cdot 1 \cdot 4 \cdot 3 \cdot 2 \cdot 1} =$ 15 possible 4-person committees

LESSON 10-9
Challenge
Roundtable Discussion

The number of ways in which 4 people can be seated *in a row*, on a bench that seats 4 is $_4P_4$, or 4!.

$\frac{4}{\text{1st seat}} \times \frac{3}{\text{2nd seat}} \times \frac{2}{\text{3rd seat}} \times \frac{1}{\text{4th seat}} = 4! = {_4P_4} = 24$ different arrangements

Now consider what happens if 4 people are seated *in a circle*,

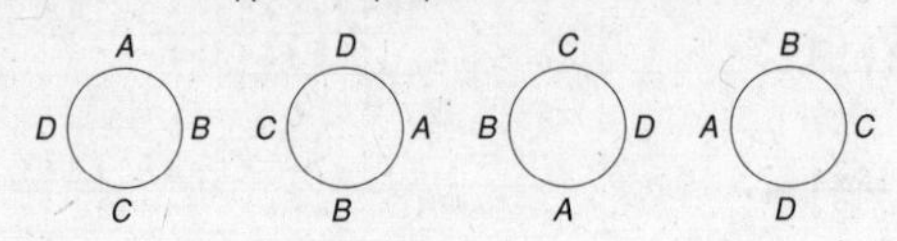

around a round table that seats 4.

Note that the 4 circular arrangements shown are really all the same with respect to who sits next to whom.

For each of the 4! permutations, there are 4 alike.

So, there are fewer ways to seat 4 people at a circular table that seats 4.

$\frac{_4P_4}{4} = \frac{4!}{4} = \frac{4 \cdot 3 \cdot 2 \cdot 1}{4} = 6$ different arrangements

1. In how many different ways can 5 people be seated in a row, on a bench that seats 5?

$_5P_5 = 5! = 5 \cdot 4 \cdot 3 \cdot 2 \cdot 1 = 120$

2. In how many different ways can 5 people be seated in a circle, around a circular table that seats 5?

$\frac{_5P_5}{5} = \frac{5!}{5} = \frac{5 \cdot 4 \cdot 3 \cdot 2 \cdot 1}{5} = 24$

3. In how many different ways can n people be seated in a row, on a bench that seats n? Answer in factorial form.

$n!$

4. In how many different ways can n people be seated in a circle, around a circular table that seats n? Answer in factorial form.

$(n-1)!$

LESSON 10-9
Problem Solving
Permutations and Combinations

Write the correct answer.

1. In a day camp, 6 children are picked to be team captains from the group of children numbered 1 through 49. How many possibilities are there for who could be the 6 captains?

13,983,816 possibilities

2. If you had to match 6 players in the correct order for most popular outfielder from a pool of professional players numbered 1 through 49, how many possibilities are there?

10,068,347,520 possibilities

Volleyball tournaments often use pool play to determine which teams will play in the semi-final and championship games. The teams are divided into different pools, and each team must play every other team in the pool. The teams with the best record in pool play advance to the final games.

3. If 12 teams are divided into 2 pools, how many games will be played in each pool?

15 games

4. If 12 teams are divided into 3 pools, how many pool play games will be played in each pool?

6 games

A word jumble game gives you a certain number of letters that you must make into a word. Choose the letter for the best answer.

5. How many possibilities are there for a jumble with 4 letters?

A 4
B 12
(C) 24
D 30

6. How many possibilities are there for a jumble with 5 letters?

F 24
G 75
(H) 120
J 150

7. How many possibilities are there for a jumble with 6 letters?

A 120
B 500
(C) 720
D 1000

8. On the Internet, a site offers a program that will un-jumble letters and give you all of the possible words that can be made with those letters. However, the program will not allow you to enter more than 7 letters due to the amount of time it would take to analyze. How many more possibilities are there with 8 letters than with 7?

F 5040
G 20,640
(H) 35,280
J 40,320

LESSON 10-9
Reading Strategies
Use a Visual Aid

A **permutation** is an arrangement of objects in a certain order.

How many different ways can you arrange these three shapes?

□ ○ △

You can use a tree diagram to visualize all of the possible arrangements:

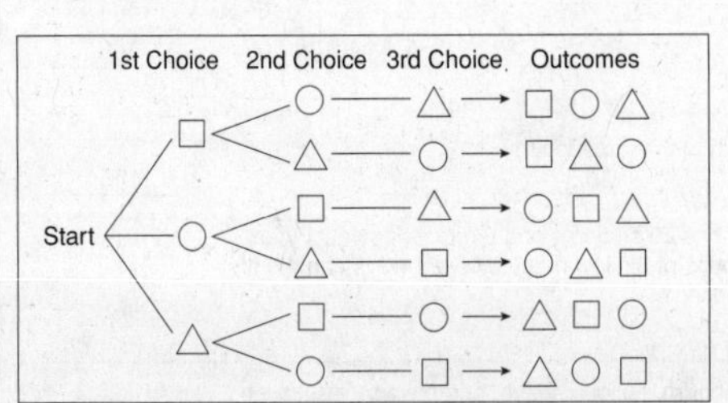

Use the tree diagram to answer the following.

1. If you start with the circle, how many different arrangements can you make? List them.

2; circle, square, triangle; circle, triangle, square

2. If you start with the square, how many different arrangements can you make? List them.

2; square, circle, triangle; square, triangle, circle

3. If you start with the triangle, how many different arrangements can you make? List them.

2; triangle, square, circle; triangle, circle, square

4. How many different arrangements can you make with these three shapes?

6

LESSON **10-9**

Puzzles, Twisters & Teasers

Finding a Treasure!

Black out the incorrect expressions to see a shape.

9! = 362,880	2! = 4	11! = 9,497,876	3! = 9	5! = 120
5! = 25	3! = 6	4! = 14	2! = 2	10! = 100
6! = 150	5! = 125	8! = 40,320	7! = 49	6! = 36
4! = 256	7! = 5040	10! = 10,000,000	10! = 3,628,8000	7! = 823,543
4! = 24	9! = 729	12! = 144	8! = 16,777,216	11! = 39,916,800

What do you see? the shape of a large X